CRITICAL THINKING & LOGICAL REASONING WORKBOOK-6

GIFT OF LOGIC™ SERIES

An Essential Resource for Everyone

Boost Your Thinking Skills

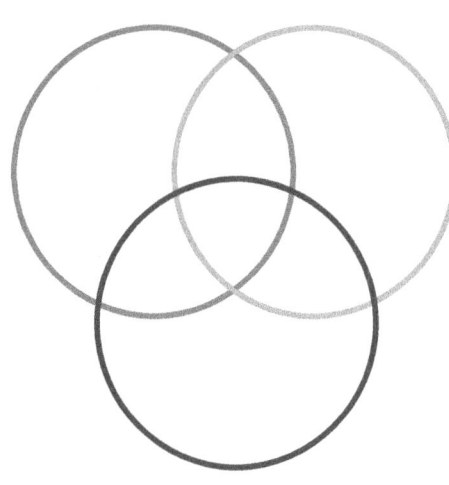

Verbal Reasoning

Analytical Reasoning

Pictorial Reasoning

THIRD EDITION

| FOR GRADES 6-12 | STUDENTS, TEACHERS, AND PARENTS |

Ranga Raghuram

GIFT OF LOGIC™

 **Gift Of Logic, Inc**
http://www.giftoflogic.com
sales@giftoflogic.com

Critical Thinking and Logical Reasoning Workbook-6
ISBN-13: 978-1494832445
ISBN-10: 1494832445

Third Edition
1-2014

Copyright © 2009 Gift Of Logic, Inc. All rights reserved. No part of this publication may be reproduced, stored in a retrieval system, transmitted in any form or by any means, electronic, mechanical, photocopying, recording or otherwise, without the written permission of the publisher.

License: This book is licensed for use by one person only. Use of this book in a group setting (classroom, workshop, etc) without the written permission of the publisher is prohibited. Unauthorized duplication is strictly prohibited by law. Contact the publisher at sales@giftoflogic.com for classroom/school/group licensing.

GIFT OF LOGIC™
CRITICAL THINKING & LOGICAL REASONING CURRICULUM
12 WORKBOOKS TO BOOST YOUR THINKING SKILLS

For Kindergarten, Grade 1, and Grade 2

Workbook# 0

Verbal Reasoning	Finding the truth, Inferencing, Analogies, Synonyms and Antonyms, Agree/Disagree
Analytic Reasoning	Memory drill, Decision making, Positioning, Sudoku
Pictorial Reasoning	Connect the dots, Mazes, Picture Sequence, Spot the difference, etc

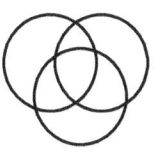
Workbook# 1

Verbal Reasoning	Finding the truth, Inferencing, Analogies, Synonyms and Antonyms, Agree/Disagree
Analytic Reasoning	Sorting, Positioning, Picking, Assorted problems, Numeric and Alphabetic Sudoku
Pictorial Reasoning	Picture Sequence, Spot the difference, Odd picture

Workbook# 2

Verbal Reasoning	Finding the truth, Classification, Direct and Inverse relationship, Inferencing, Analogies, Agree/Disagree
Analytic Reasoning	Sequencing, Scheduling, Strategy, Picking, etc
Pictorial Reasoning	Picture Analogy, Odd picture, Pattern matching, etc

For Grade 3, Grade 4, and Grade 5

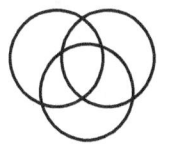
Workbook# 3

Verbal Reasoning	Not, And, Or, If .. then, Conditional inferencing, Unconditional inferencing, Symbolic Logic
Analytic Reasoning	Lists, Sequencing, Grouping, Venn Diagrams, Graph logic, Number logic, Letter logic, Sudoku
Pictorial Reasoning	Picture sequence, Picture analogy, Odd picture, Picture difference, Pattern matching

Workbook# 4

Verbal Reasoning	Contradiction, Converse, Inverse, Contrapositive, Conditional inferencing, Symbolic Logic
Analytic Reasoning	Scheduling, Looping, FIFO, LIFO, Correlation, Venn Diagram, Graph logic, Number logic, Sudoku, etc
Pictorial Reasoning	Picture sequence, Picture analogy, Odd picture, Picture difference, Pattern matching

Workbook# 5

Verbal Reasoning	Biconditional, Categorical inferencing, Cause and Effect, Symbolic Logic, Agree/Disagree, Word and Sentence analogy
Analytic Reasoning	Correlation, Grouping, Venn Diagrams, Graph logic, Number logic, Letter logic, Sudoku, etc
Pictorial Reasoning	Picture sequence, Picture analogy, Odd picture, Picture difference, Pattern matching

********* Essential resource for everyone *********
*http://www.giftoflogic.com   *sales@giftoflogic.com

GIFT OF LOGIC™
CRITICAL THINKING & LOGICAL REASONING CURRICULUM
12 WORKBOOKS TO BOOST YOUR THINKING SKILLS

For Grades 6-12, College/University Students, Adults

Primer

Prereq

Verbal Reasoning	Logical Operators, Conditional, Categorical and Causal reasoning, Validity, Fallacies, Symbolic Logic
Analytic Reasoning	Positioning, Grouping, Sudoku
Pictorial Reasoning	Pattern perception, Figure formation, Paper folding and cutting, Figure matrix, Rule detection

Workbook# 6

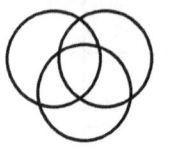

Verbal Reasoning	Arguments-Main point, Must be true, Cannot be true
Analytic Reasoning	Positioning, Grouping, Sudoku
Pictorial Reasoning	Pattern perception, Figure formation, Paper folding and cutting, Figure matrix, Rule detection

Workbook# 7

Verbal Reasoning	Arguments-Strengthening, Weakening
Analytic Reasoning	Positioning, Grouping, Sudoku
Pictorial Reasoning	Pattern perception, Figure formation, Paper folding and cutting, Figure matrix, Rule detection

Workbook# 8

Verbal Reasoning	Arguments - Controversy, Paradox
Analytic Reasoning	Positioning, Grouping, Sudoku
Pictorial Reasoning	Pattern perception, Figure formation, Paper folding and cutting, Figure matrix, Rule detection

Workbook# 9

Verbal Reasoning	Arguments- Assumptions, Reasoning strategy
Analytic Reasoning	Positioning, Grouping, Sudoku
Pictorial Reasoning	Pattern perception, Figure formation, Paper folding and cutting, Figure matrix, Rule detection

Workbook# 10

Verbal Reasoning	Arguments-Flawed reasoning, Analogous reasoning
Analytic Reasoning	Positioning, Grouping, Sudoku
Pictorial Reasoning	Pattern perception, Figure formation, Paper folding and cutting, Figure matrix, Rule detection

********* Essential resource for everyone *********
Get the GIFT OF LOGIC™ today !
*http://www.giftoflogic.com *sales@giftoflogic.com

Dear Reader:

Your decision to purchase this book is commendable. You now have in your hands, a comprehensive, easy-to-read book in Critical thinking and Logical reasoning that will introduce you to three different areas of thinking and reasoning - Verbal, Analytical and Pictorial. Solving problems in Verbal Reasoning is important to develop a critical mind. Solving problems in Analytic Reasoning is important to develop a flexible and resourceful mind. Solving problems in Pictorial Reasoning is important to develop a visually alert mind.

This book is presented in a workbook format to help you progress quickly. Parents and teachers are urged to complete the exercises ahead of the student and assist them whenever necessary with the help of detailed answers provided at the end of the book. This book can be used as a supplementary resource in the regular class room or it can be used during winter and summer vacations. College/University students, working professionals and retired individuals will also find the Gift Of Logic(tm) Series very useful in enhancing their problem solving abilities, confidence and general intellect.

Critical thinking and Logical reasoning must be practiced consistently to develop strong cognitive skills. After completing the exercises in this book, continue to read the other books in this series to get familiar with different types of Logical reasoning problems.

This workbook is one in a series of twelve workbooks. Please refer to the brochure before this page for a brief description of each workbook. Visit the website http://www.giftoflogic.com for more information.

Happy thinking and reasoning!

TABLE OF CONTENTS

Verbal Reasoning

Main Point..8
Must be true..23
Cannot be true..40

Analytical Reasoning

Sudoku..59
Positioning..62
Grouping...70

Pictorial Reasoning

Patter perception..76
Figure formation...78
Paper folding and cutting...80
Figure matrix..81
Rule detection..83

Answers

Verbal..86
Analytic...134
Pictorial..147

Certificate of Completion

Name —————————————— Date ——————————

VERBAL REASONING

Name _____   Date _____

MAIN POINT (CONCLUSION)

In this section on "Main Point", you will develop the ability to identify the main point of an argument.

A complete argument is given (premises and conclusion) and your task is to identify the conclusion and pick a statement that matches the conclusion of the argument.

The main point of an argument is its conclusion. After the argument is presented, questions such as the following will be posed.
  *Which one of the following is the main point of the argument?
  *Which one of the following matches the conclusion of the argument?
  *The main point of the argument is that..
  *The argument made by the politician is that..

The correct answer is the one that rephrases the conclusion accurately. Statements that restate the premises are incorrect. Statements that do not match the conclusion accurately and completely, or those that are entirely different from the conclusion are also incorrect.

When answering main point questions, read the argument carefully, looking for the statement that expresses the conclusion of the argument. Conclusion keywords like "so", "therefore", "hence" and "thus" give important clues to identify the conclusion in the argument. Then, try to find the answer choice that closely matches this conclusion.

| 1 | MAIN POINT | Cars |

Hybrid cars are more efficient than cars that run on one fuel source alone. They will help keep our environment clean. Therefore, we must support the use of hybrid cars.

The main point of the argument is that

A) hybrid cars run more miles than non-hybrid cars with the same amount of fuel.
B) hybrid cars emit less pollutants than non-hybrid cars.
C) we must not hesitate to drive hybrid cars.

| 2 | MAIN POINT | Birds |

Birds have a sound dictionary in their brain that they use to communicate with each other. This dictionary tells them what sounds to make in situations such as when they are lost or when they are hungry. This leads us to believe that birds use their sound dictionary in the same way we human beings use our word dictionary.

The main point of the argument is that

A) when birds do not know how to express themselves, they refer to their sound dictionary.
B) birds are not intelligent enough to use a sound dictionary.
C) without the sound dictionary, birds do not know how to communicate.

3 — MAIN POINT — Earth

The astronauts brought several pounds of rocks from the Moon. The chemical analysis of these rocks indicate that they are mainly composed of silicon, aluminum and calcium which are also found on some rocks on Earth. This similarity, combined with the fact that some of the Moon rocks were estimated to be as old as Earth itself, leads us to surmise that the Earth and Moon have similar origins.

Which of the following represents the main point of this argument?

A) Moon rocks behave the same way as rocks found on Earth in chemical laboratory tests.
B) Moon rocks are estimated to be as old as those of terrestrial rocks.
C) Moon and Earth are thought to have formed at the same time.

4	MAIN POINT	Cars

That some cars are very expensive and some cars are moderately priced is a known fact. It is also known that cars are traditionally considered as a status symbol in society – the more expensive the car one drives, the more popular one becomes in his or her social circle. However, these days, one can see some very rich people driving moderately priced cars and some not so rich people driving very expensive cars, thereby challenging the value of cars as a status symbol.

The conclusion of the above argument is matched by which one of the following?

A) driving more expensive cars will make one appear to be more fashionable in his or her friend's circle.
B) the price of very expensive cars has come down in recent months.
C) the position of cars as a status symbol is in doubt.

5   MAIN POINT   Planets

Scientists have long believed that Pluto is the last planet in our Solar System. But recently, thanks to new facts that were not available before, a committee of world scientists have decided that Pluto does not qualify to be a planet at all, let alone being the last planet. Therefore, it should come as no surprise to anyone, if after several years, based on new facts about our solar system that are not available today, any other planet that we know today is also disqualified by the committee.

The main point of the argument is that

A) Pluto was considered the last planet in the Solar System for hundreds of years.
B) any planet faces the risk of being disqualified from our Solar System if new information is available to the committee.
C) Pluto does not orbit around the sun.

6	MAIN POINT	Accounting

Government Official: There are different types of taxes - Income tax, Sales tax, Property tax, etc. These taxes give the government the revenues needed to run its business. Income taxes contribute a large share to the government treasury. It is illegal for anyone to hide their income from the government because if they did so, then the government will not be able to build roads and dams, provide health care, etc. Thus, everyone must report their income accurately to the government.

The government official argues that

A) it is legal if property information is not reported accurately.
B) if government does not build roads and dams, it is because people have not reported their income accurately.
C) reporting false income information to the government is not acceptable.

| 7 | MAIN POINT | Medical |

Some children pester their parents to get one ice-cream cone every week. Their parents succumb to this pestering because they do not see any harm in doing so. But, some parents hold the view that eating more than one ice-cream cone a month is unhealthy and refuse to give in to pestering by their children. This goes to prove that not every parent agrees on the number of ice-cream cones that their children can have every month.

The main point of this argument is that

A) some parents like their children to have ice-cream cones whereas some parents do not.
B) children who get one ice-cream cone a month do not pester their parents to buy more.
C) some parents disagree over how many ice-cream cones a child can have in a month.

| **8** | MAIN POINT | Pilots |

Women certainly can fly airplanes as well as men. In fact, women fly both commercial and military airplanes and have received gold medals for their flying skills. Arguments that incorrectly imply that women cannot pilot airplanes as well as men arise mainly because of the fact that only a small percentage of pilots are women.

The main point of this argument is that

A) because only a small percentage of pilots are women, they are not able to pilot planes as well as men.
B) women and men fly airplanes the same way.
C) women do not have sufficient skills to fly airplanes.

| 9 | MAIN POINT | Sports |

People who participate in professional sports face the risk of acute pain and injuries caused because of excessive stress to their muscles and bones. The urgent need of these people to get well quickly and return back to their sports was not adequately met by the medical profession. Consequently, a new occupation emerged, called Sports Medicine, with the purpose of treating and preventing sports related injuries. Comprising mainly of doctors, nurses, physiotherapists and chiropractors, sports medicine is a lucrative profession for those interested in helping people involved in sports.

The main point of this paragraph is that

A) people who are injured must get treated by a Sports Medicine doctor.
B) a sports medicine doctor is different from a regular doctor.
C) the new occupation called "sports medicine" was created to address injuries specific to sportsmen.

10	MAIN POINT	Tools

Modern woodworking tools have become popular because they are powered by electricity. The electric drill is a popular tool to drill holes and drive screws inside wood. Electric tools also do polishing, sanding and a variety of other woodworking tasks. They are available in different sizes ranging from small sizes for home use to big machines for use at the lumber yard. At the same time, older tools such as the chisel and the plane have not lost their popularity among carpenters and are used even today. This goes to show that people use tools regardless of whether they are modern or not.

The main point in the above argument is that

A) if a tool is old, it will not be preferred.
B) certain older tools have lost their relevance in modern day woodworking.
C) age is not a factor that decides if a tool is used or not.

| 11 | MAIN POINT | Automobiles |

Automobile manufacturers are enticing customers to buy more powerful vehicles by reducing the price on these vehicles. It may really be a lot of fun to drive powerful vehicles, but they consume more fuel than normal vehicles. This will increase the cost of maintaining them. Therefore, even though they may be affordable, they are not cheap to own over the long run.

The main point of the above argument is that

A) the total cost of owning a powerful vehicle over a long time is high.
B) the vehicles with a lot of power are exciting to drive.
C) an increase in fuel cost will increase the cost of maintenance.

| 12 | MAIN POINT | Geology |

People living near the foothills of mountains face the danger of volcanic eruptions that can spew different types of materials. The most dangerous volcanic material called the Lava, is basically hot molten rocks flowing from the mouth of the volcano down the slopes of the mountain. But, despite this danger, people living near the foothills have frequently ignored calls by authorities to evacuate their homes during times of eruption. Hence, their refusal to evacuate is sufficient proof of their extraordinary bravery.

The main point of this argument is that

A) people living in foothill must obey evacuation orders.
B) most people have defied evacuation orders of authorities at times of eruption.
C) people who defy volcano related safety warnings have an abnormal level of braveness.

| 13 | MAIN POINT | College |

Even before going to college, Abdul already knew everything about electricity. He acquired this knowledge from his father, who is an Electrical Engineer. But, a college degree is a prerequisite for getting good jobs in any profession. So, he decided to get a college degree in Electrical Engineering.

The main point of the passage above is that

A) since Abdul knew everything about electricity, he does not need a college degree in Electrical Engineering.
B) Abdul decided to get a college degree in Electrical Engineering in order to get a good job.
C) Abdul can use his father's degree in Electrical Engineering to get a good job.

14 MAIN POINT — Politics

Politician:
    I want to give tax breaks to industries to attract more jobs to our city. Simultaneously, I want to give job training to all unemployed people. The job training program is crucial because without it, when the tax breaks attract jobs to our city, there will not be any city worker capable of doing these jobs.

The main point of the politician's speech is that

A) it is crucial for unemployed workers to acquire job training.
B) tax breaks are crucial to attract more jobs to the city.
C) all the city workers are currently unemployed.

MUST BE TRUE (CORRECT CONCLUSION)

In this section on "Must be true", you will develop the ability to identify inferences that must be true.

One or more premises are given, but the conclusion is not given. You must apply the inferencing methods that you have learned in the book titled "Critical thinking & Logical Reasoning Primer" and pick an answer choice that must be true if the given premises are true. In other words, you must pick the correct conclusion that can be deduced (inferred) from the premises.

After the premises are presented, questions such as the following will be posed.

* If the above facts are true, which one of the following must be true?
* Which one of the following is a correct conclusion that can be inferred?
* The facts presented most strongly support which one of the following?
* Which one of the following can be inferred from the premises?

The correct answer is one that we can infer from the premises to be true. This correct answer would be a logical conclusion of the argument.

Incorrect answers are those that cannot be true or those that are vague or out of context. Incorrect answers will not be a logical conclusion of the argument.

Note that you can encounter conditional statements, categorical statements, causal statements, etc in the argument.

Name _____ Date _____

1	MUST BE TRUE	Identity

Jack and Jill went up the hill. Jack fell down and Jill came tumbling down the hill.

If the above information is true, then which one of the following must be true?

A) Jack came tumbling down the hill.
B) Jill fell down.
C) Jack broke his crown.

2	MUST BE TRUE	Conditional

If the temperature becomes very warm, an ozone-alert will be issued. It is very warm today with a temperature of 100 degrees Fahrenheit.

If the above statements are true, then which one of the following must also be true?

A) An ozone-alert is likely to be issued today.
B) An ozone-alert will be issued today.

| 3 | MUST BE TRUE | All/some |

Every item in this shop is foreign. All foreign items are expensive.

If the above premises are correct, then which one of the following conclusions can be drawn?

A) Not every foreign item in this shop is expensive.
B) Every item in this shop is cheap.
C) Every item in this shop is expensive.

| **4** | MUST BE TRUE | Except |

Doctors at the Regional Hospital must give preference to children over adults except when an adult has a serious medical problem and the child does not.

If the above facts are true, then which one of the following must be true?

A) If Donna takes her daughter to the hospital, her daughter will be treated before any other adult who is waiting for treatment.
B) Roger, an adult with a serious injury was treated before a child with runny nose.
C) Alfred's infant son will be treated after a woman who has been waiting for ten minutes is treated for a headache.

| 5 | MUST BE TRUE | Including |

All dogs bark when they perceive danger, including situations such as when something is thrown at them. The other day, Jill's dog saw a stranger walking nearby and sensed a dangerous situation.

Based on the above information, which one of the following can be inferred?

A) Jill's dog barked at the stranger.
B) Jill' dog did not bark at the stranger.

| 6 | MUST BE TRUE | Before/After |

Before the year 2004, all the students of Mount Everest Middle School wore a white shirt. After 2004, the school policy was changed and students were required to wear a blue shirt. Mansoor, Emily and Silvia have been students of Mount Everest Middle School since 2002.

If the above facts are true, then which one of the following must be true?

A) Before 2004, Mansoor wore a green shirt to school.
B) In 2004, Emily wore a white shirt to school.
C) In 2005, Silvia wore a blue shirt to school.

| 7 | MUST BE TRUE | However |

The Columbia Hospital has an emergency wing and a non-emergency wing. The emergency wing is open 24 hours a day. The non-emergency wing is open from 8 AM to 5 PM only. However, the non-emergency wing must receive patients from the emergency wing if the doctors there are very busy. One day, at 9 AM, there was a bus accident near the hospital and the doctors at the emergency wing were overwhelmed with patients.

Which one of the following can be inferred from the above?

A) On the day of the bus accident, the doctors at the emergency wing attended to patients requiring non-emergency treatment.
B) On the day of the bus accident, the doctors at the non-emergency wing attended to patients requiring urgent treatment.

| 8 | MUST BE TRUE | Astrology |

For a long period of time, several human beings have strongly believed in Astrology. Their belief in Astrology is so strong that when the stars and planets are not positioned in the sky to their liking, they postpone the completion of important tasks.

Which one of the following can be inferred from the information presented above?

A) Most of us believe in Astrology.
B) If important tasks got completed, then the stars and planets are not positioned to the satisfaction of people who hold a strong belief in Astrology.
C) If important tasks got completed in time, then the stars and planets are positioned to the satisfaction of people who hold a strong belief in Astrology.

| 9 | MUST BE TRUE | Jurassic |

The Jurassic age, when dinosaurs roamed the earth, is considered to be from 238 million years ago to 138 million years ago. The dinosaurs became extinct after this period. No evidence has been discovered so far regarding their life before this period.

The statements above, if true, will help us to reach which one of the following conclusions?

A) Dinosaurs roamed the earth 135 million years ago.
B) Dinosaurs roamed the earth since 240 million years ago.
C) Dinosaurs roamed the earth 150 million years ago.

| **10** | MUST BE TRUE | Limits |

If you purchase something with a credit card, you can pay later when the bill arrives, but you can purchase only up to your credit limit. But, if you make a purchase with a debit card, you must have enough money available in your bank account.

If the above facts are true, then which one of the following must be true?

A) If one wants to purchase a TV for 500 dollars, but has only 50 dollars in the bank, a credit card can be used.
B) If one wants to purchase a watch for 50 dollars, and has only 50 dollars in the bank, a debit card can be used.
C) If one's credit limit is 1000 dollars, he can buy a TV for 1500 dollars using a credit card.

| **11** | MUST BE TRUE | Classification |

Birds can be broadly classified as either carnivorous or herbivorous. Carnivorous birds mainly eat animal food, but also eat plant food. Herbivorous birds eat plant food only.

The above premises support which one of the following?

A) Herbivorous birds can eat the food that carnivorous birds eat.
B) Carnivorous birds can eat the food that herbivorous birds eat.
C) There is no food that both carnivorous and herbivorous birds can eat.

| 12 | MUST BE TRUE | Cholesterol |

Cholesterol will increase the viscosity of blood, thereby making it difficult for the blood to flow smoothly in the body. Dairy products and oil have high cholesterol content in them.

Which one of the following can be inferred from the paragraph?

A) Dairy products and oil will impede the flow of blood.
B) Dairy products and oil will assist in the smooth flow of blood.

| 13 | MUST BE TRUE | Government |

It is unethical to seek favors of any kind from government officials. But, it is not unethical to seek favors from people who are not government officials. This is because, government officials work for the benefit of public and their salaries are paid with public money, whereas this is not the case with people who are not government officials.

Which one of the following can be inferred from the above?

A) It is ethical to seek financial favors from friends who are government officials.
B) It is not ethical to seek financial favors from relatives who are not government officials.
C) It is ethical to seek favors from friends who are not government officials.

| **14** | MUST BE TRUE | Nevertheless |

Daniel prepared very well for the Logic exam. Nevertheless, he was not able to score as well as Rachel. It has been proven that one's score in the Logic exam is directly proportional to the number of hours of sleep one gets the night before the test and that no other factor can influence the score.

The statements above, if true, can lead to which one of the following conclusions?

A) Rachel is more intelligent than Daniel.
B) Daniel is more intelligent than Rachel.
C) Rachel slept more hours than Daniel the night before the exam.
D) Daniel slept more hours than Rachel the night before the exam.

| 15 | MUST BE TRUE | Notwithstanding |

Notwithstanding the fact that Bob is an experienced sportsman, he cannot be included in the team because of his bad behavior. Bad behavior, particularly by sportsmen, do not set a good example to youngsters.

Which one of the following can be inferred from the statement above?

A) If a sportsman is very talented he can be included in the team even if his behavior is bad.
B) Bad behavior by people other than sportsmen and sportswomen is acceptable to the younger generation.
C) Behavior that does not set a good example to youngsters will disqualify a talented sportsman from the team.

| 16 | MUST BE TRUE | In spite of |

In spite of repeated warnings to comply with traffic rules, Zac continues to drive above speed limits. Today, he was stopped by traffic police for speeding. The penalty for a speeding offense is expensive, but the penalty for a non-speeding offense is not.

Which one of the following can be inferred from the passage?

A) Zac has to pay an inexpensive fine.
B) Zac has to pay an expensive fine.
C) Zac does not owe any fine.

| 17 | MUST BE TRUE | Identity |

Few students from Hopkins Middle School and a few students from Preston Middle School were selected for a trip to meet the astronauts at the Space Center. Martha and Steve enjoyed their meeting with the astronauts. Steve wore his Preston school shirt on the trip. Anne, a student of Hopkins Middle was not selected for the trip.

Which one of the following conclusions can be made from the passage?

A) Martha and Steve are students of Preston Middle School.
B) Only two students were selected for the trip to the Space Center.
C) Steve and Anne do not go to the same school.

CANNOT BE TRUE (INCORRECT CONCLUSION)

In this section on "Cannot be true", you will develop the ability to identify inferences that cannot be true. This ability will give you the confidence to spot arguments that have invalid inferences.

One or more premises are given, but the conclusion is not given. After the premises are presented, questions are posed as follows:

*Which one of the following cannot be true?
*Which one of the following does not follow?
*Which one of the following is an incorrect conclusion that is inferred from the facts presented above?
* If the above information is true, then which one of the following must be false?

The correct answer is one that cannot be inferred. That is, the correct answer is the one that cannot be true if the given premises are true.

Incorrect answers are those that must be true and can be inferred.

As you evaluate each answer choice, think of the phrase "cannot be true" for each answer choice. If you find a answer choice that cannot be true, then that choice is the correct answer. If the answer choice is a "must be true" or a "can be true" answer choice, then it is not the correct answer.

Verbal Reasoning
© Gift Of Logic, Inc * Copying prohibited

| 1 | CANNOT BE TRUE | Win and Lose |

The Sharks team played five soccer games against the Panthers team. A game can be won or lost, but it cannot end in a tie.

If the above facts are true, which one of the following cannot be true?

A) The Sharks team won five games.
B) The Sharks team lost three games and the Panthers team lost two games.
C) The Sharks team won two games and lost two games against the Panthers.

| 2 | CANNOT BE TRUE | Conditional |

Students in a school must choose courses from a list that includes Algebra, Trigonometry and Calculus. If they take Algebra, they must also take Trigonometry. Students who take Trigonometry must also take Calculus.

If the above conditions are true, then which one of the following cannot be true regarding Jennifer, a student of the school?

A) She takes Algebra, Trigonometry, and Calculus.
B) She takes Trigonometry and Calculus, but not Algebra.
C) She takes Algebra and Calculus, but not Trigonometry.

| 3 | CANNOT BE TRUE | Contrapositive |

Calcium is very important for the bones because it makes the bones become strong. If sufficient calcium is not present in the diet, then the bone mass will go below the normal level.

If the above information is true, which one of the following must be false?

A) If a person does not have sufficient calcium in his diet, he will have bone mass below the normal level.
B) If a person does not have below normal level of bone mass, then he does not have sufficient calcium in his diet.
C) If a person has a normal bone mass level, he will have sufficient calcium in his diet.

| 4 | CANNOT BE TRUE | Cars |

A car dealer sold 100 cars during the first month and a total of 1000 cars in 2006. He sold 900 cars during the first month and a total of 1000 cars in 2007.

If the above facts are true, which one of the following cannot be true?

A) The dealer sold more cars in March 2006 than he did in March 2007.
B) Excluding the first month, more cars were sold in 2007 than in 2006.
C) The dealer sold the same number of cars during both years.

| 5 | CANNOT BE TRUE | Categorical |

P, Q and R are the names of three groups of people. No one belongs to both P and Q. No one belongs to both Q and R.

If the above facts are true, which one of the following cannot be true?

A) No one belongs to P, Q and R.
B) No one belongs to both P and R.

| 6 | CANNOT BE TRUE | Physics |

To create a vacuum in a given area, air inside the area is pumped out and the area is sealed. Since there are no air molecules in vacuum, energy cannot flow through it. If the area is not sealed, outside air will rush in to fill the vacuum. Several devices, such as the thermos flask, use the concept of vacuum to provide insulation against transfer of energy.

If the above statements are true, which one of the following must be false?

A) In a thermos flask with an unbroken vacuum, hot water will remain hot and cold water will remain cold.
B) A thermos flask whose vacuum is broken will retain energy as efficiently as a thermos flask with an unbroken vacuum.

| 7 | CANNOT BE TRUE | Buildings |

There are five buildings of type A and ten buildings of type B. Buildings of type B can be converted to buildings of type C by painting them blue, and buildings of type B can be converted to buildings of type A by painting them green.

If the above statements are true, which one of the following must be false?

A) After applying the blue paint on half the buildings of type B, there are the same number of buildings of each type.
B) There is at least one building of type C after the green paint is applied to half the buildings of type B.

| 8 | CANNOT BE TRUE | Teaching |

In order to teach Physics at any university, one must have a Ph.D in Physics. Mr. Rogers teaches Physics at the Everest middle school.

Which one of the following is an incorrect conclusion derived from the above facts?

A) Mr. Rogers has a Ph.D in Physics.
B) If you do not have a Ph.D in Physics, you cannot teach Physics at any University.

| 9 | CANNOT BE TRUE | Biology |

Mold, a type of fungus, is a black or green hairy mass that grows when its spores settle down on a surface suitable for reproduction. Common places where you can find it are in the surfaces of fruits and vegetables, bread, and in living plants and organisms. It penetrates into edible food and makes it poisonous. It can grow inside the human body and make people fall sick. It is destructive in nature and the substances that it infects are of no use to mankind. However, it performs an essential function in our ecosystem by breaking down organic matter.

If the information in the above paragraph is true, which one of the following cannot be true?

A) Fungus can be found inside the human body.
B) Since fungus is destructive in nature, it is of no use to mankind.
C) When fungus settles on organic matter, the matter disintegrates into its components.

| 10 | CANNOT BE TRUE | Medical |

Gastric juice is a fluid created in the stomach. The food that we eat is broken down when it mixes with this fluid. Gastric juice is made up of hydrochloric acid, pepsin and mucin. Pepsin in an enzyme that breaks the food with the help of the acid. Mucin is a part of the mucus, which protects the wall of the stomach from acid. This protection is critical, without which peptic ulcer is caused, which in turn results in heartburns and other problems. The erosion of stomach walls can happen due to high levels of hydrochloric acid in the stomach, smoking and stress.

If the information presented above is true, which one of the following must be false?

A) Smoking is the only cause of ulcer in stomach.
B) Hydrochloric acid is required to break down the food in the stomach.

| 11 | CANNOT BE TRUE | Survey |

Many women in China like to assume names that signify fragrance. For example, "Lifen", meaning "fragrance" in Chinese, is a name used by many women. Surveys have shown that, at least twenty percent of women in each major city have a name that is related to fragrance in some way.

If the above statements are true, which one of the following must be false?

A) "Lanfen", which in Chinese means "orchid fragrance" is a name that some Chinese women would not like to be called with.
B) Women in China do not like to have their name associated with anything fragrant.

| **12** | CANNOT BE TRUE | Sports |

The number "3" is a lucky number for the Sharks soccer team. If one of their players has this number in his jersey, then they will win the game.

If the statement above is true, which one of the following cannot be true?

A) A player with jersey# 3 did not play and yet Sharks won a game.
B) Sharks did not win the game even though jersey# 3 played.
C) The Panthers had a player with jersey# 3 and they won the game.

| 13 | CANNOT BE TRUE | Science |

Dr. Henry liked solitude. As a scientist, he performed all his experiments himself without anyone's help. Not only that, he also avoided meetings and social functions except when they were of importance to his scientific interests. He made several stunning scientific discoveries in Physics and Chemistry.

If the above information is true, then which one of the following cannot be true?

A) Dr. Henry's discoveries in Physics and Chemistry were performed when no one except himself was present.
B) Dr. Henry's team made several scientific discoveries in Physics and Chemistry.
C) Dr. Henry made outstanding discoveries in Physics and Chemistry during his time.

| 14 | CANNOT BE TRUE | Percentage |

The Stanley Middle School starts sharp at 8 AM. Students who come late to school are marked as being tardy. Last year, out of 500 students, 50 of them got a tardy mark. But, in the previous year, 100 students out of a total of 200 were marked as tardy.

If the above statements are true, then which one of the following cannot be true?

A) More students were tardy to school during the previous year when compared to last year.
B) A higher percentage of students were tardy during the previous year when compared to last year.
C) There were more punctual students during the previous year when compared to last year.

Verbal Reasoning
© Gift Of Logic, Inc * Copying prohibited

| 15 | CANNOT BE TRUE | Health |

Everyone knows how to handle a headache when it occurs, but there is no permanent cure for this. Headaches are said to occur mainly due to stress in the muscles and nerves. Most headaches can be relieved by using off-the-shelf analgesics, but some headaches, such as the ones caused by migraine are treated by using prescription medicine.

Which one of the following does not follow from the facts presented above?

A) Off-the-shelf analgesics can get rid of all types of headaches.
B) Headaches caused by migraine cannot be relieved by off-the-shelf analgesics.

16 — CANNOT BE TRUE — Charity

Employees of Alpha Construction Company have set up a disaster relief fund to collect money for use by them after a catastrophic event. The money collected will be put in a bank account and withdrawn when there is a need. Since there could be disagreement among individuals regarding the distribution of funds, the disaster relief committee has formulated detailed rules. Included in the rules is a statement that only the employees who contribute to the fund can be beneficiaries.

Which one of the following does not follow from the facts presented above?

A) When a disaster strikes the city where the company is located, the employees can claim benefits from the disaster relief fund.
B) When a disaster strikes the city where the company is located, anyone can claim benefits from the fund.
C) If you are not a beneficiary, then you did not contribute to the fund.

| 17 | CANNOT BE TRUE | Management |

Calls to the Customer Service department of the Coolride Biking Company are put on hold for approximately thirty minutes. This is because of the fact that they are short handed due to the resignations of five employees several days ago. Customers have recently complained about this inordinate delay to the management of the company. The management listened and rectified the situation immediately.

If the information in the above passage is true, then which one of the following does not follow?

A) The delays have worsened since the complaints were made.
B) The delays have improved since the complaints were made.

ANALYTICAL REASONING

Name _____   Date _____

1
SUDOKU

Solve the following Sudoku. A correctly solved Sudoku has numbers 1-9 appearing only once in each row, each column and each 3x3 grid. Solving Sudokus will help you to gain valuable analytic skills.

	1	9		3	7		2	5
		5	2		6	8		9
7	2		5	4		1	3	
4	3		1	9		5	6	
1		6	7		2	9		3
	9	2		6	4		8	1
2	7		6	8		3	9	
9		3	4		1	2		8
	4	5		2	3		1	7

Analytical Reasoning   Answers-134   59
© Gift Of Logic, Inc * Copying prohibited

SUDOKU

Solve the following Sudoku. A correctly solved Sudoku has numbers 1-9 appearing only once in each row, each column and each 3x3 grid. Solving Sudokus will help you to gain valuable analytic skills.

	7	2	8	1	6	5	9	4
6		9	3		5	8		2
1	8		4	9		3	6	
	2	6	7	8	3	9	4	1
4		8	5		1	6		3
7	3		6	4		2	8	
	6	4	9	5	7	1	3	8
9		7	2		8	4		6
8	5		1	6		7	2	

SUDOKU

Solve the following Sudoku. A correctly solved Sudoku has numbers 1-9 appearing only once in each row, each column and each 3x3 grid. Solving Sudokus will help you to gain valuable analytic skills.

1		5	7		4	3		2
6	2		8	5		9	7	
9	7	4	2	3	6	5	8	1
2		8	3		7	6		5
3	5		6	2		1	4	
7	6	1	9	4	5	2	3	8
8	1		4	6		7	5	
	3	7		8	9		2	6
4	9	6	5	7	2	8	1	3

Analytical Reasoning

1		
	POSITIONING	constraint

In the grid shown below, there are three cells marked with an X. In the empty cells, write the letter Y so that a X is not below a Y.

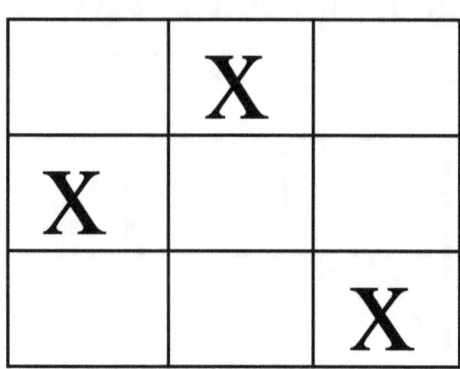

1) How many Y can you place in the empty cells?

2		
	POSITIONING	constraint

In the grid shown below, there are four cells that have a square in them. In the empty cells, draw a circle so that there is no square to its left.

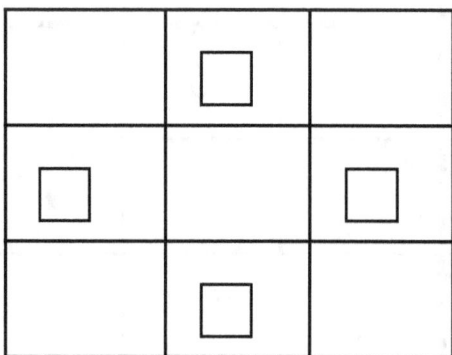

1) How many circles can you place in the empty cells?

| 3 | POSITIONING | except |

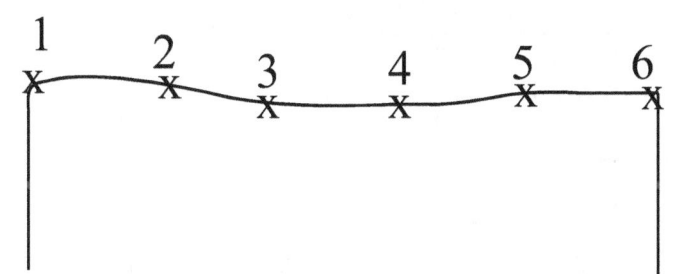

A string is tied to two poles as shown. A bird can sit anywhere in the six marked positions except on even numbered positions.

1) Can the bird sit on the pole 1?

2) In how many positions can the bird sit?

| 4 | POSITIONING | or, only |

There are several slots where you can fit either a 60 Watt bulb or a 40 Watt bulb as shown. The 60 Watt slots can hold either a 60 Watt bulb or a 40 Watt bulb. A 40 Watt slot can hold only a 40 Watt bulb. Answer the questions below.

1) A 40 Watt slot can hold a 60 Watt bulb.
    A) True   B) False
2) The number of slots where you can fit a 60 Watt bulb is
    A) 3   B) 5
3) The number of slots where you can fit a 40 Watt bulb is
    A) 3   B) 5

5 POSITIONING move

7	8 E	9
4	5	6
1 H	2	3

The grid above shows nine numbered cells. The H represents a horse and E represents an Elephant. The Horse can move a total of three spaces from its current position, with two spaces in a different direction than the third.

The sequence of moves that the Horse can make to reach the Elephant is
    A) (1,2,5,6)
    B) (1,2,5,8) or (1,4,7,8)
    C) (1,4,5,8)

6 POSITIONING move

7	8	9
4	5	6 M
1	2 H	3

The grid above shows nine numbered cells. The H represents a horse and the M represents a monkey. The Horse can move a total of three spaces from its current position, with two spaces in a different direction than the third.

The sequence of moves that the Horse can make to reach the monkey is
    A) (2,5,6)
    B) (2,5,8,9)
    C) (2,1,4,7) and (7,8,9,6)

Name ——————————— Date ———————————

7   POSITIONING   constraint

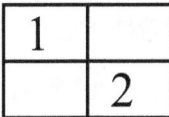

The grid above shows four cells with two of them filled with numbers.

Is it possible to fill the blank cells with numbers 1 and 2 so that each row and each column has two different numbers?

    A) Yes   B) No

8   POSITIONING   fill

3		
	2	
		1

The grid above has some cells already filled in.

Fill the blank cells with numbers 1, 2, and 3 so that each row and each column has the numbers 1, 2, and 3.

9   POSITIONING   fill

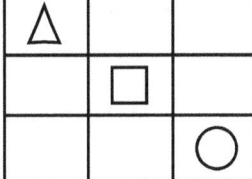

The grid above has some cells already filled in.
Fill the blank rows with symbols △, ○, and □ so that each row and each column has one of each shape.

Analytical Reasoning
© Gift Of Logic, Inc * Copying prohibited

Name _____   Date _____

10   POSITIONING   skip

There are nine train stations, numbered 1 to 9. Train A starts at station# 1 and stops at odd numbered stations. Train B starts at station# 1 and skips two stations between stops. Hansel boarded train A at station# 1 and Gretel boarded train B at station# 1.

At which station should they get down so that they can meet each other?
    A) station# 6
    B) station# 7
    C) station# 9

11   POSITIONING   fill

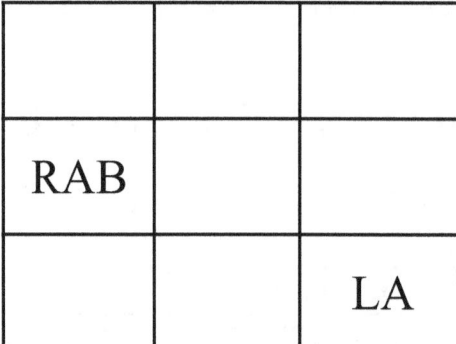

In each cell, write L, B, A, or R depending on whether each cell has a neighbor to the Left, Below, Above or Right respectively. For example, RAB means that the cell has a neighbor to the Right, Above and Below. Similarly, LA means that the cell has a neighbor to the Left and Above. L for Left, R for Right, A for Above, and B for below. Fill in the other cells.

12 POSITIONING rules

Place the letters A, B, C, D, E, F, G, and H in the squares below based on the following rules.

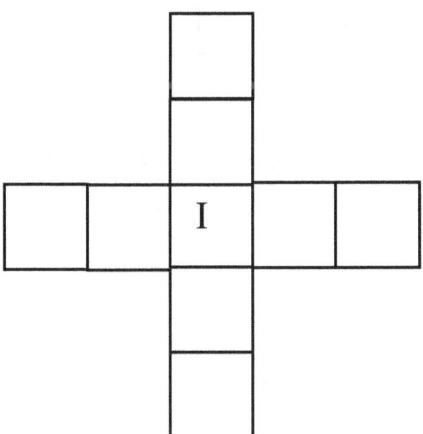

A must be above B.
I must not be above A.
C must not be next to I.
D must be immediately to the right of C.
H and F must be at the extreme ends.
G must be before H.

13 POSITIONING with rules

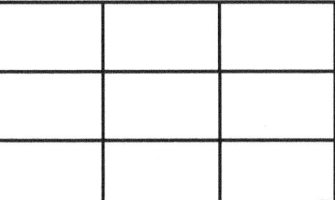

Place the letters A, B, C, D, E, F, G, H, and I in the grid above. The letter A must be in the first cell (top-left) and must be immediately above B. C must be immediately to the right of B. F must be to the right of C, below G, and immediately above E. I must be below B and D must be immediately after I.

Name _____   Date _____

14 POSITIONING   move

A	B	C
D	E	F
G	H	I

What is the new position of the letters A, B, C, D, E, F, G, H, and I after they are moved around in the order of the following rules. Fill the grid below with the new positions of the letters.

    The letters in columns 1 and 3 are interchanged.
    The letters in rows 1 and 3 are interchanged.

15 POSITIONING   move

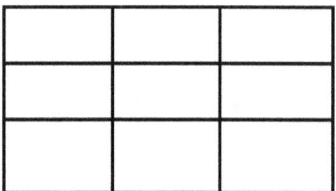

What is the new position of the letters after they are moved one spot clockwise? Fill the blank circle with the new positions.

Analytical Reasoning
© Gift Of Logic, Inc * Copying prohibited

Name _____   Date _____

16 POSITIONING — at most

A		
2		C
	B	

The empty cells in the grid shown above are to be filled with numbers and letters so that there are at most three numbers.

1. What is the maximum number of letters that the grid can hold?

17 POSITIONING — together

SCENARIO

Red, Green, Blue, Orange, and Yellow balls are to be placed on a table one after the other in five spots numbered 1,2,3,4, and 5. Red ball must be the first ball. Green and Blue balls must be together. Orange ball must be immediately before the green ball. Yellow ball can be placed before or after the orange ball.

Answer the questions below.

1) Green ball can be the third ball.
    A) True   B) False

2) Yellow ball can be the third ball.
    A) True   B) False

| 1 | GROUPING | cannot |

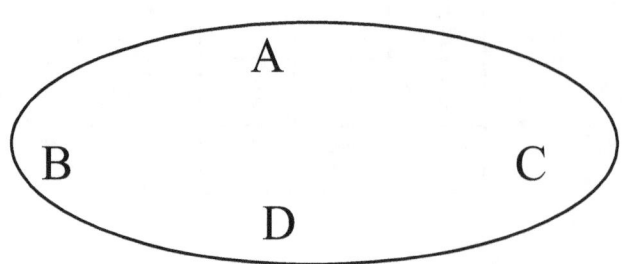

A tray contains alphabets A, B, C, and D.
Pick two alphabets from the tray.
If A is picked, D cannot be picked.

What are the possible picks?   Write them below.

| 2 | GROUPING | must |

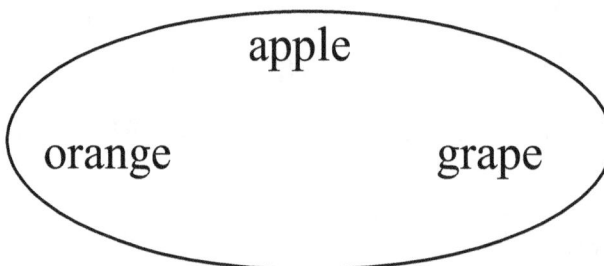

Pick any two fruits from the above tray. If apple is picked, then grape must also be picked.

What are the possible picks?   Write them below.

Name _____ Date _____

| 3 | GROUPING | selective |

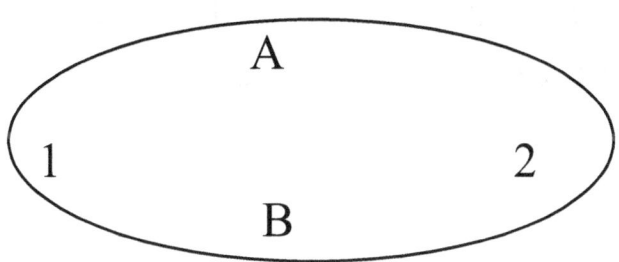

Select one alphabet and one number from the tray.

What are the possible picks?   Write them below.

| 4 | GROUPING | at least, at most |

Science - 5 books
Math - 5 books
History - 5 books

There are five Science books, five Math books, and five History books in three shelves as shown above.

You must select at least 2 Science books, at most 2 Math books, and at least 3 history books.

1) What is the maximum number of books that can be selected?

2) What is the minimum number of books that can be selected?

| 5 | GROUPING | multiple groups |

Team P has four members and team Q has four members as shown. Pick two from team P and two from team Q. Member A cannot be selected with E or F or both.

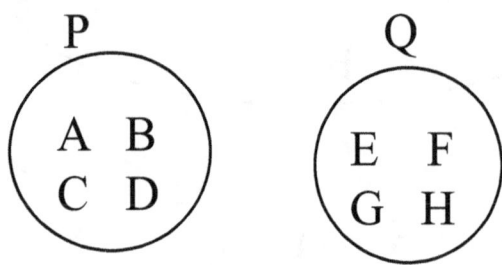

1) Which of the following selections are valid?

```
A)  A   B   G   F
B)  E   H   A   D
C)  C   B   E   F
```

| 6 | GROUPING | multiple groups |

Team A has four members (A,B,C,D) and team B has four members (E,F,G,H). Pick two from team A and two from team B. If A is selected, E must be selected. If C is selected, G must be selected.

1) Which of the following selections are valid?

```
A)  A   B   F   G
B)  C   A   G   E
C)  E   G   B   D
```

Name _____   Date _____

7   GROUPING   regrouping

Class A
| 7 boys 4 girls |

Class B
| 6 boys 3 girls |

Class A has 7 boys and 4 girls. Class B has 6 boys and 3 girls.

To make the classes to be of same size, what must be done?

  A) Move 1 boy from class A to class B.
  B) Move 1 boy and 1 girl from class A to class B.

8   GROUPING   regrouping

Box-1
| 7 squares
3 triangles |

Box-2
| 3 squares
5 triangles |

Box-1 contains 7 squares and 3 triangles. Box-2 contains 3 squares and 5 triangles.

Regroup the squares and triangles, by moving some of the shapes from one box to another, so that both the groups have the same number of triangles and squares.

How many squares and rectangles are there in the regrouped boxes?

Analytical Reasoning   Answers-146

| **9** | GROUPING | less |

From a group of 5 cars and 5 trucks, select a group of 5 vehicles so that there are less number of cars than trucks. There should be at least one vehicle of each type in the selected group.

How many groups can be selected?

| **10** | GROUPING | more |

From a basket of 5 oranges and 5 apples, select a group of five fruits so that there are more apples than oranges. There should at least be one fruit of each type.

How many groups can be selected?

Analytical Reasoning

PICTORIAL REASONING

Name _____   Date _____

PATTERN PERCEPTION - MISSING PATTERN

Find the correct figure from the three alternatives given, that will fit logically into the missing portion of the figure on the left.

1 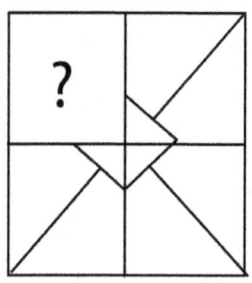   A 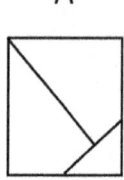   B 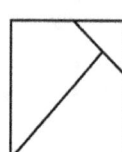   C

2 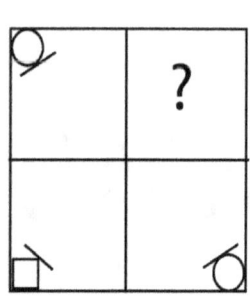   A 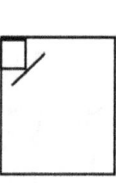   B 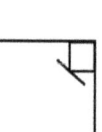   C

3 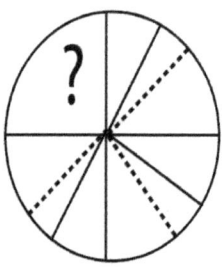   A    B    C

4 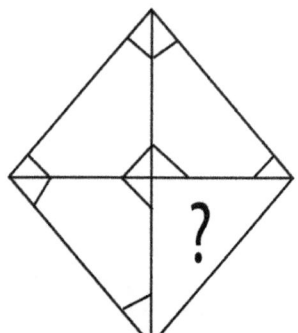   A 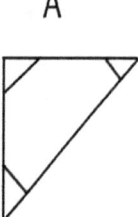   B 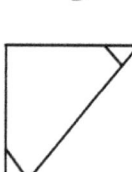   C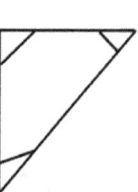

Pictorial Reasoning   Answers-147
© Gift Of Logic, Inc * Copying prohibited

Name ——————————— Date ——————

PATTERN PERCEPTION - CONTINUING PATTERN

Find the correct figure from the two alternatives given, that will logically continue the pattern of figures on the left.

5

   X X X
X X
X   ?   A    B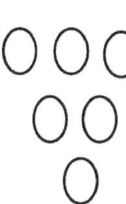

6

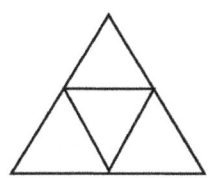   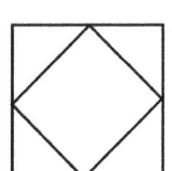   ?   A 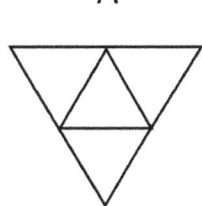   B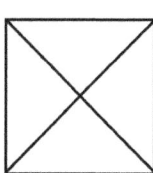

7

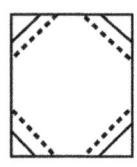   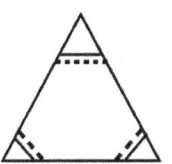   ?   A 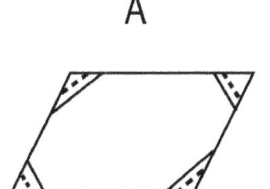   B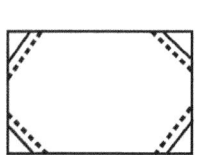

8

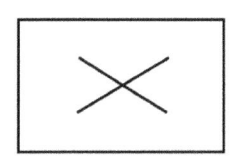      ?   A 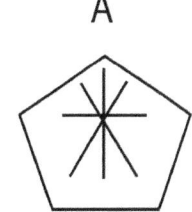   B

Pictorial Reasoning
© Gift Of Logic, Inc * Copying prohibited

Name _____   Date_____

FIGURE FORMATION

Find the correct figure that will be formed, when the two figures on the left are joined together.

                                        A   B

1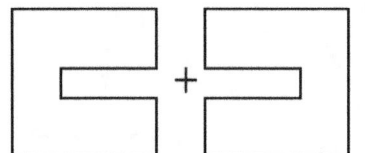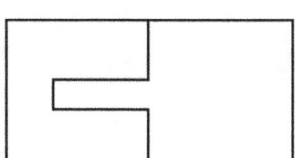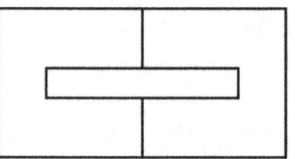

                                        A   B

2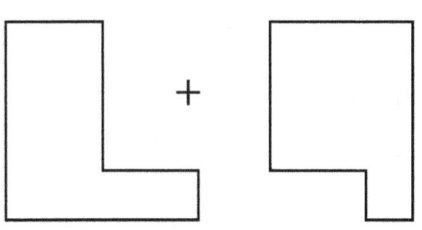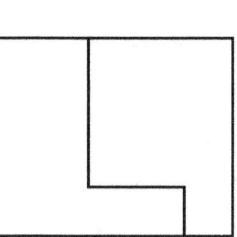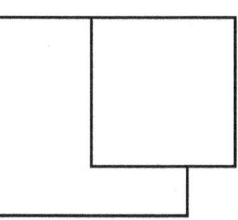

                                      A   B

3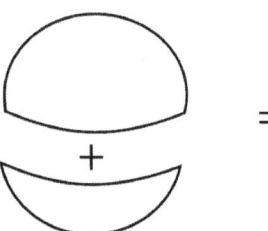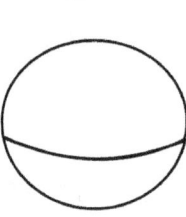

                                A   B

4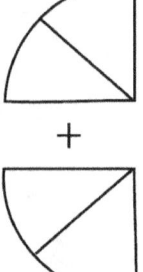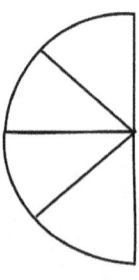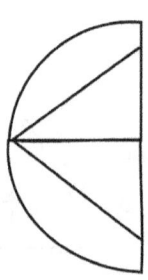

Pictorial Reasoning

Name _____   Date _____

FIGURE FORMATION

Find the correct figure that will be formed, when the two figures on the left are joined together.

                                              A   B

5 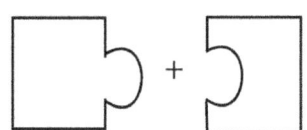 =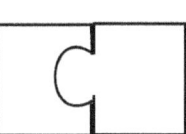

                                              A   B

6 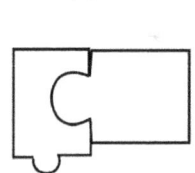 =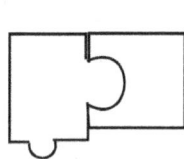

                                              A   B

7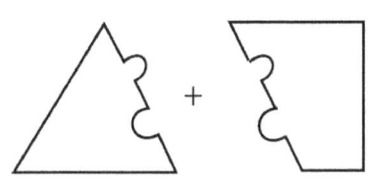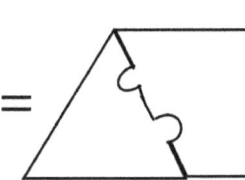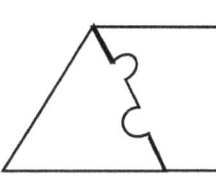

Pictorial Reasoning   Answers-147
© Gift Of Logic, Inc * Copying prohibited

Name ——————————————————   Date ———————————————

PAPER FOLDING AND CUTTING

Find the correct figure that will be formed when the paper on the left is folded in the direction of the arrow, and then holes are cut in it as shown.

1

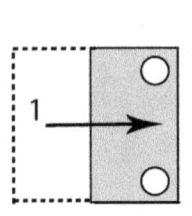   A B C D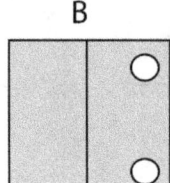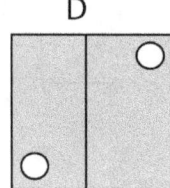

2

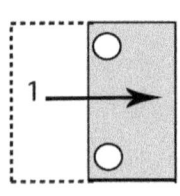   A B C D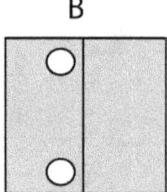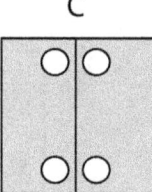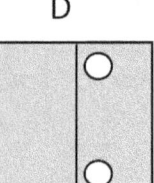

3

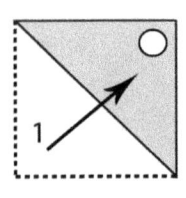   A B C D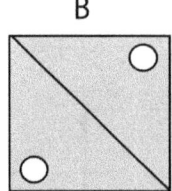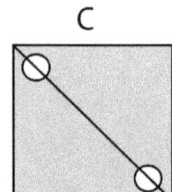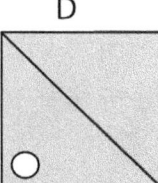

Pictorial Reasoning
© Gift Of Logic, Inc * Copying prohibited

FIGURE MATRIX- SIMILARITY

Three figures in the 2 x 2 matrix have similar characteristics. Find the fourth figure from the alternatives given that is also alike.

1 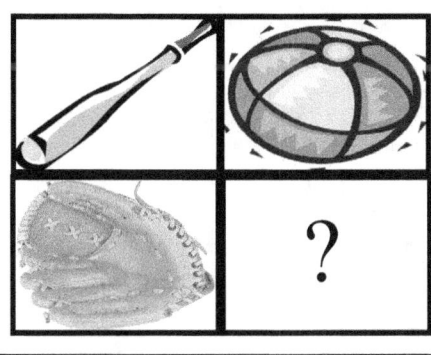 A  B  C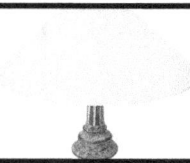

2 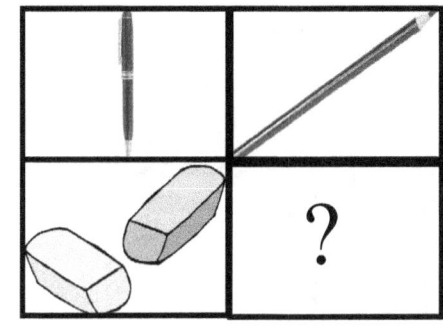 A  B  C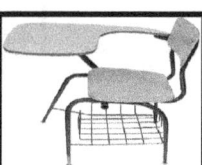

3  A  B  C

4  A  B  C

Name _____   Date _____

FIGURE MATRIX- ANALOGY

Find the correct figure from the alternatives given that will fit in the empty box, such that the bottom two figures are related in the same way as the top two figures.

5    A    B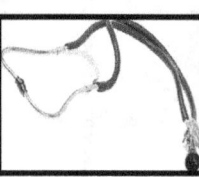

6    A    B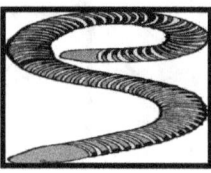

7 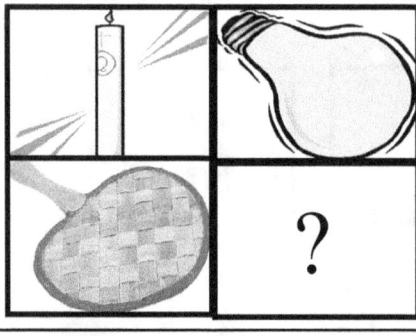   A    B

8    A    B

Pictorial Reasoning
© Gift Of Logic, Inc * Copying prohibited

RULE DETECTION

Read the given rule in each question. Then, find the correct choice from the alternatives given that satisfies the rule.

1. The shaded circle moves clockwise

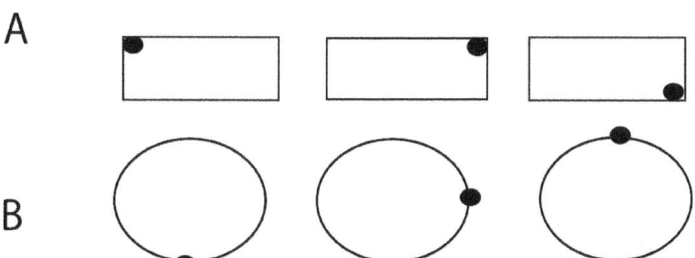

2. The arrows rotate clockwise

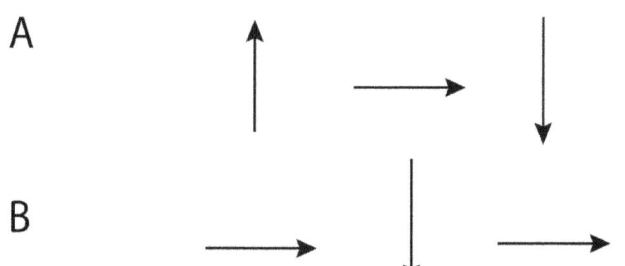

3. The outside figures move anticlockwise and the inside figures move clockwise

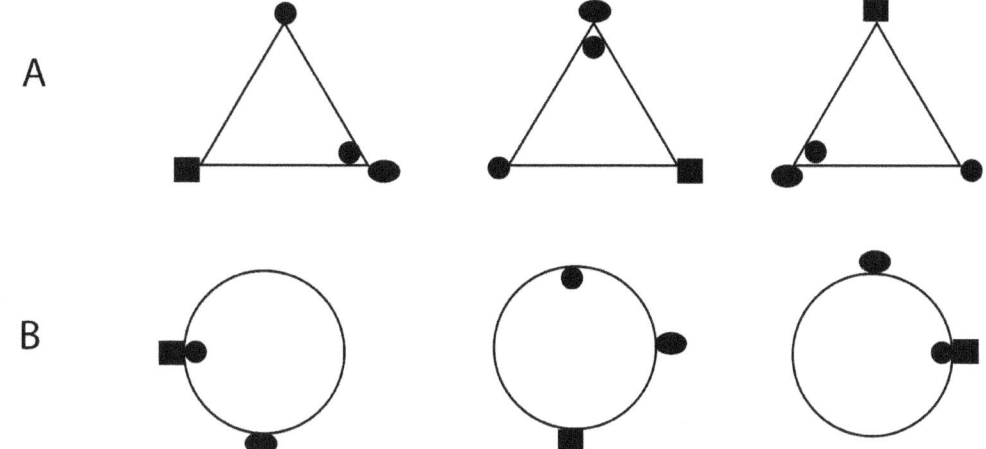

Name —————————————— Date ——————————————

RULE DETECTION

Read the rule in each question. Then, find the correct choice from the alternatives given that satisfies the rule.

4 Number of objects inside the box increases from left to right

A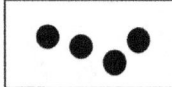

B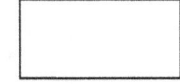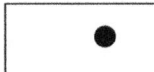

5 Objects become bigger as the sequence progresses

A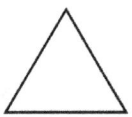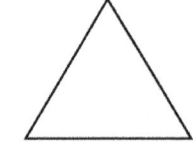

B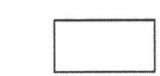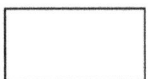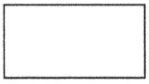

6 Objects become smaller as the sequence progresses

A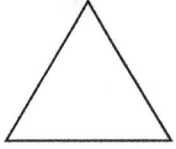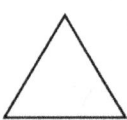

B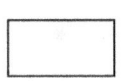

Pictorial Reasoning

ANSWERS

1   MAIN POINT

Hybrid cars are more efficient..

The main point of the argument is that
A) hybrid cars run more miles than non-hybrid cars with the same amount of fuel.
B) hybrid cars emit less pollutants than non-hybrid cars.
C) we must not hesitate to drive hybrid cars.

ANSWER

Answer: C

The word "therefore" in the last statement indicates that this is the conclusion. The conclusion of the argument is "we must support the use of hybrid cars". This is the main point of the argument.

A - incorrect – this could be true and maybe why the author says that hybrid cars are efficient. This statement provides an additional fact to support the conclusion, but this is not the main point.

B - incorrect – perhaps hybrid cars emit less pollutants than non-hybrid cars and that is the reason why they can be utilized to help keep our environment clean. This statement is an additional fact that was not stated in the argument, but it is not the main point.

C - correct – "we must not hesitate to drive hybrid cars" is a rephrase of the conclusion that "we must support the use of hybrid cars" and so, this is the correct answer.

Answers
© Gift Of Logic, Inc * Copying prohibited

2 MAIN POINT

Birds have a sound dictionary in their brain that..

The main point of the argument is that
A) when birds do not know how to express themselves, they refer to their sound dictionary.
B) birds are not intelligent enough to use a sound dictionary.
C) without the sound dictionary, birds do not know how to communicate.

ANSWER

Answer: A

The words "this leads us to" indicates the conclusion of the argument namely "birds use their sound dictionary in the same way that we human beings use our word dictionary". We can now find the answer choice that matches this conclusion.

A - correct – this statement matches the conclusion. This choice states that when the birds do not know how to express themselves, they refer to their sound dictionary. This is similar to the conclusion that states that birds use their sound dictionary the same way we use our word dictionary.

B - incorrect – this statement is contrary to the premise that "birds have sound dictionary in their brain that they use to communicate".

C - incorrect - this statement does not match the conclusion in meaning. We cannot infer this from the facts presented in the argument.

Answers
© Gift Of Logic, Inc * Copying prohibited

3	MAIN POINT

The astronauts brought several pounds..

Which one of the following represents the main point of this argument?
A) Moon rocks behave the same way as rocks found on Earth in chemical laboratory tests.
B) Moon rocks are estimated to be as old as those of terrestrial rocks.
C) Moon and Earth are thought to have formed at the same time.

ANSWER

Answer: C

Conclusion of this argument is "..leads us to surmise that the Earth and Moon have similar origins".

A - incorrect – how the rocks from Moon and Earth behave in chemical laboratory tests is not the main point.

B - incorrect – this is a rephrase of the premise, but not the conclusion.

C - correct – passage says ".. leads us to surmise that Earth and Moon have similar origins" – which is the same as saying that Moon and Earth are thought (surmised) to have formed at the same time. The word "origin" in the conclusion means "starting point".

4   MAIN POINT

That some cars are very expensive and some cars..

The conclusion of the above argument is matched by which one of the following?
A) driving more expensive cars will make one appear to be more fashionable in his or her friend's circle.
B) the price of very expensive cars has come down in recent months.
C) the position of cars as a status symbol is in doubt.

ANSWER

Answer: C

The conclusion of this argument is ".. thereby challenging the value of cars as a status symbol."

A - incorrect – this statement is a rephrase of the premise. The premise states that "the more expensive the car one drives the more popular one becomes" – this choice says the same thing – to be fashionable is the same as being popular.

B - incorrect – this is not stated in the argument and is also not the conclusion of the argument.

C - correct – this is the same as saying that the value of cars as a status symbol is being challenged. This is the main point of the argument.

> **5** MAIN POINT

Scientists have long believed that Pluto is the last planet ..

The main point of the argument is that
A) Pluto was considered the last planet in the Solar System for hundreds of years.
B) any planet faces the risk of being disqualified from our Solar System if new information is available to the committee.
C) Pluto does not orbit around the sun.

ANSWER

Answer: B

The conclusion (main point) of the argument is that no one should be surprised if any other planet is disqualified to be a planet by the committee if new facts are found about our solar system.

A - incorrect – this is a restatement of the premise.

B - correct – the conclusion of the argument is that it should not come as a surprise if any other planet is disqualified because of new information – this choice says that any planet faces the risk of being disqualified from our solar system based on new facts. So, this choice matches the conclusion and is the main point.

C - incorrect – The facts in the argument do not say that Pluto does not orbit around the Sun. The argument does not mention why Pluto does not qualify to be a planet. This statement does not match the conclusion.

Answers
© Gift Of Logic, Inc * Copying prohibited

6	MAIN POINT

Government Official: There are different types of taxes...

The government official argues that
A) it is legal if property information is not reported accurately.
B) if government does not build roads and dams, it is because people have not reported their income accurately.
C) reporting false income information to the government is not acceptable.

ANSWER

Answer: C

The conclusion is "everyone must report their income accurately to the government". Note that the question is posed as "the government official argues that". We argue to prove a (main) point. We argue to justify our conclusion. So, this question is the same as the others that ask you to identify the main point.

A - incorrect – the main point is that everyone must report their income accurately - not whether it is legal or not to report their property information accurately.

B - incorrect – in the passage, the official says that if people hide their income, then government cannot build roads. This choice, however states the converse of the conditional statement, which is not true.

C - correct - this choice says that reporting false income information to the government is not acceptable. This is the same as the conclusion and is the main point that the government official argues about.

Answers
© Gift Of Logic, Inc * Copying prohibited

7   MAIN POINT

Some children pester their parents to get one ice-cream ..

The main point of this argument is that
A) some parents like their children to have ice-cream cones whereas some parents do not.
B) children who get one ice-cream cone a month do not pester their parents to buy more.
C) some parents disagree over how many ice-cream cones a child can have in a month.

ANSWER

Answer: C

The conclusion of this argument is "this goes to prove that not every parent agrees on how many ice-cream cones their children can have every month".

A - incorrect - the main point is not whether parents like their children to eat ice-cream cones or not. The main point is the disagreement over how many ice-cream cones their children can have.

B - incorrect - this statement is in contradiction to the stated premise that some parents "refuse" to give in to pestering and buy only one ice-cream cone during a month.

C - correct - this statement matches the conclusion. It is the disagreement over how many ice-cream cones their kids can eat in one month that is the main point.

8 MAIN POINT

Women certainly can fly airplanes as well as men.

The main point of this argument is that
A) because only a small percentage of pilots are women, they are not able to pilot planes as well as men.
B) women and men fly airplanes the same way.
C) women do not have sufficient skills to fly airplanes.

ANSWER

Answer: B

The conclusion is that "women certainly can fly airplanes as well as men". Facts are provided to support this conclusion. The passage also states that there are arguments available that incorrectly imply the negation of this conclusion. The conclusion of this passage is a bit difficult to spot. The the word "certainly" indicates the presence of a conclusion.

A - incorrect - this statement is referred to as an incorrect implication.

B - correct - this statement rephrases the conclusion. Since the conclusion is that women can fly airplanes as well as men, we can conclude that they fly airplanes the same way as men.

C - incorrect - this choice talks about women not having sufficient skills to fly airplanes. This is not stated and is not the conclusion of the argument.

9	MAIN POINT

People who participate in professional sports face..

The main point of this paragraph is that
A) people who are injured must get treated by a Sports Medicine doctor.
B) a sports medicine doctor is different from a regular doctor.
C) the new occupation called "sports medicine" was created to address injuries specific to sportsmen.

ANSWER

Answer: C

The conclusion of the paragraph is "consequently, a new occupation emerged called Sports Medicine..". The use of the word "consequently" signifies that the sentence has a conclusion. The conclusion is not stated at the beginning or end of the argument. This is not the normal structure of an argument, but nevertheless, its conclusion can be identified.

A - incorrect – The argument does not say that anyone with injuries must see a Sports Medicine doctor. Those doctors mainly treat sport-related injuries. This is not the main point.

B - incorrect – this is not the main point.

C - correct - this is the main point that paraphrases the conclusion accurately.

10 MAIN POINT

Modern wood working tools have become popular..

The main point in the above argument is that
A) if a tool is old, it will not be preferred.
B) certain older tools have lost their relevance in modern day woodworking.
C) age is not a factor that decides if a tool is used or not.

ANSWER

Answer: C

The argument presents premises to explain how modern woodworking tools powered by electricity have become popular. Then, it goes on to say that older tools have not lost their popularity either. Finally, it concludes that people use tools regardless of whether they are modern or not.

A - incorrect - this choice does not match the conclusion. It is contrary to the conclusion.

B - incorrect - there is no reference in the argument to older tools losing their relevance.

C - correct - this is the main point of the argument. This choice says that age is not a factor that decides if a tool is used or not. This is the same as the conclusion of the argument.

11 MAIN POINT

Automobile manufacturers are enticing customers..

The main point of the above argument is that
A) the total cost of owning a powerful vehicle over a long time is high.
B) the vehicles with a lot of power are exciting to drive.
C) an increase in fuel cost will increase the cost of maintenance.

ANSWER

Answer: A

The conclusion (main point) of the argument is that even though powerful vehicles may be affordable, they are not cheap to own over the long run.

A - correct – this statement is similar to the main point. The total cost of owning a powerful vehicle over a long time is high because of the fuel costs associated with its high fuel consumption.

B - incorrect – this is not the main point, but a stated premise.

C - incorrect – this is a stated premise, not the main point.

12	MAIN POINT

People living near the foothills of some mountains..

The main point of this argument is that
A) people living in foothill must obey evacuation orders.
B) most people have defied evacuation orders of authorities at times of eruption.
C) people who defy volcano related safety warnings have an abnormal level of braveness.

ANSWER

Answer: C

The conclusion is that the refusal of people to evacuate is sufficient proof of their extraordinary bravery.

A - incorrect – this statement does not match the conclusion.

B - incorrect – this choice presents an incorrect fact and is not the main point of the argument.

C - correct – the conclusion of the argument is that the refusal by people to evacuate is sufficient proof of their extraordinary bravery. This choice echoes the conclusion by saying that people who defy (refuse) the volcano related safety warnings have an abnormal (extraordinary) level of braveness. So, this choice is the main point of the argument.

13	MAIN POINT	

Even before going to college, Abdul already knew..

The main point of the passage above is that
A) since Abdul knew everything about electricity, he does not need a college degree in Electrical Engineering.
B) Abdul decided to get a college degree in Electrical Engineering in order to get a good job.
C) Abdul can use his father's degree in Electrical Engineering to get a good job.

ANSWER

Answer: B

The conclusion is that he decided to get a college degree in Electrical Engineering.

A - incorrect – this is not the main point. This choice asserts something that cannot be found in the argument.

B - correct – this choice matches the conclusion and is the main point.

C - incorrect – this choice makes an assertion that is not mentioned in the argument.

| **14** | MAIN POINT |

Politician:
   I want to give tax breaks to industries..

The main point of the politician's speech is that
A) it is crucial for unemployed workers to acquire job training.
B) tax breaks are crucial to attract more jobs to the city.
C) all the city workers are currently unemployed.

ANSWER

Answer: A

Though there is no clear conclusion indicator, we can identify the conclusion if we arrange the information logically as follows:

   I want to give tax breaks to industries to attract jobs.
   I also want to give job training to unemployed.
   When tax breaks attract jobs to our city there will be no trained
    workers to do the job.
   Therefore, the job training is crucial. (this is the conclusion/main point)

A - correct – this is the main point as it matches the conclusion.

B - incorrect - this is not the main point. This is one premise that is stated in order to lead us to the main point.

C - incorrect - this is not the main point - moreover, the facts in the argument do not indicate that "all" the workers are unemployed.

1	MUST BE TRUE

Jack and Jill went up..

If the above information is true then which one of the following must be true ?
A)   Jack came tumbling down the hill.
B)   Jill fell down.
C)   Jack broke his crown.

ANSWER

Answer: B

Note that Jack only fell down. He did not come tumbling down. It was Jill who came tumbling down. Tumbling means "to fall and roll over suddenly".

A - incorrect - this cannot be inferred from the premises.

B - correct - since Jill came tumbling down, we can infer that she would have fallen as well.

C - incorrect - we cannot infer this from the given premises.

2 MUST BE TRUE

If the temperature becomes warm, an ozone-alert..

If the above statements are true, then which one of the following must also be true?
A) An ozone-alert is likely to be issued today.
B) An ozone-alert will be issued today.

ANSWER

Answer: B

The premise is a conditional statement that can be represented as follows:
  temperature becomes very warm → ozone alert will be issued

From this premise, we can infer that if the temperature *is* very warm, then the ozone alert will be issued.

A - incorrect - this choice uses the word "likely. This does not indicate certainty of getting an ozone alert.

B - correct - the second premise says that it is very warm today. Therefore, since the conditional premise is true, we can infer that an ozone alert will be issued today. The reasoning is shown below.

  temperature becomes very warm → ozone alert will be issued
  temperature is very warm today
  therefore, ozone alert will be issued today

If the antecedent is true, then the consequent will also be true.

3	MUST BE TRUE

Every item in this shop is foreign. All foreign items are expensive.

If the above premises are correct, then which one of the following conclusions can be drawn?
A)   Not every foreign item in this shop is expensive.
B)   Every item in this shop is cheap.
C)   Every item in this shop is expensive.

ANSWER

Answer: C

Since every item in the shop is foreign, (item → foreign) and since all foreign items are expensive, we can conclude that every item is expensive just by plain common sense reasoning.

Another way to check the answer is by drawing a Venn diagram as shown below. Draw the premises first and see which of the choices must be true.

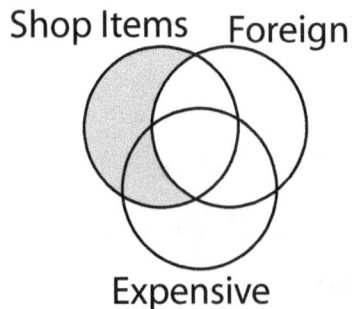

All shop items are foreign

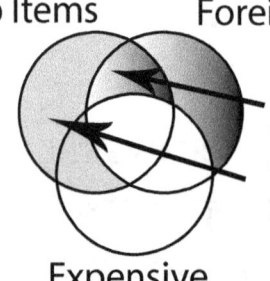

All shop items are expensive - conclusion drawn automatically after premises are drawn

All shop items are foreign
All foreign items are expensive

Note that the premise, "Every item in this shop is foreign", is the same as "All shop items are foreign". We can infer from looking at the area indicated by the arrow that "all shop items are expensive". This is the same as saying "every item in this shop is expensive".

Answers   102
© Gift Of Logic, Inc * Copying prohibited

4	MUST BE TRUE

Doctors at the Regional Hospital must give..

If the above facts are true, then which one of the following must be true?
A) If Donna takes her daughter to the hospital, her daughter will be treated before any other adult who is waiting for treatment.
B) Roger, an adult with a serious injury was treated before a child with runny nose.
C) Alfred's infant son will be treated after a woman who has been waiting for ten minutes is treated for a headache.

ANSWER

Answer: B
    child preferred over adults normally.
    if (adult is serious, but child is not) → adult preferred

Writing the facts as shown above will help to answer the questions.

A - incorrect - this depends on whether Donna's daughter is a child or not. All daughters are not children. Even if Donna's daughter was a child, whether she will be treated before any other adult depends on whether the adult has a serious problem or not.

B - correct - Roger is an adult and has a serious injury and the child has only a runny nose. So, according to the condition, Roger would be treated first.

C - incorrect - Since the woman (implies adult) has a headache and since headache is not considered serious, Alfred's infant son (implies child) would be treated before the woman is treated. So, this cannot be true.

Answers
© Gift Of Logic, Inc * Copying prohibited

5	MUST BE TRUE

All dogs bark when they perceive a danger..

Based on the above information, which one of the following can be inferred?
A)   Jill's dog barked at the stranger.
B)   Jill' dog did not bark at the stranger.

ANSWER

Answer: C

Note that the "when" in the premise is the same as "if".

perceive danger → all dogs bark
Jill's dog sensed a dangerous situation when a stranger walked nearby.
Therefore, Jill's dog will bark at the stranger.

A - correct - this can be inferred from the given premises.

B - incorrect - this cannot be true. Since the antecedent is true, the consequent must also be true. This choice says that the consequent is not true.

Answers

© Gift Of Logic, Inc * Copying prohibited

| 6 | MUST BE TRUE |

Before year 2004, all the students..

If the above information is correct, which one of the following must be true?
A) Before 2004, Mansoor wore a green shirt to school.
B) In 2004, Emily wore a white shirt to school.
C) In 2005, Silvia wore a blue shirt to school.

| ANSWER |

Answer: C
    before 2004, all students wore white shirt
    after 2004, all students wore blue shirt

A - incorrect - before 2004, all students wore a white shirt, not a green shirt.

B - incorrect - read the premises carefully - it does not say what the shirt color was for the year 2004! So, we cannot conclude that this must be true.

C - correct - 2005 is after 2004 and the shirt color after 2004 is blue. So, it must be true that Silvia wore a blue shirt.

Answers

7 MUST BE TRUE

The Columbia Hospital has an emergency..

Which one of the following can be inferred from the above?
A) On the day of the bus accident, the doctors at the emergency wing attended to patients requiring non-emergency treatment.
B) On the day of the bus accident, the doctors at the non-emergency wing attended to patients requiring urgent treatment.

ANSWER

Answer: B

Writing the facts will help us to make the correct inference.

emergency - 24 hours   non-emergency => 8-5
 if emergency doctors are very busy → non-emergency must take patients
 * emergency wing doctors were overwhelmed (on the day of accident)
 * therefore, non-emergency wing must receive emergency patients

A - incorrect - premises only say that people with emergency can be seen in the non-emergency wing, but it does not say whether people with a non-emergency condition can be seen by doctors at the emergency wing.

B - correct - we know from premises that on the day of the accident, the emergency doctors were overwhelmed. So, we can infer that emergency patients (those requiring urgent treatment) were attended by doctors in the non-emergency wing.

| 8 | MUST BE TRUE |

For a long period of time, several human beings..

Which one of the following can be properly inferred from the information presented above?
A) Most of us believe in Astrology.
B) If important tasks got completed, then the stars and planets are not positioned to the satisfaction of people who hold a strong belief in Astrology.
C) If important tasks got completed in time, then the stars and planets are positioned to the satisfaction of people who hold a strong belief in Astrology.

ANSWER

Answer: C
  ~ stars and planets are positioned to their liking → postpone tasks
  ~ postpone tasks → stars and planets are positioned to their liking
                     (contrapositive)
~ postpone tasks = completing tasks in time.
A - incorrect - premises do not say clearly that most of us believe in Astrology. It only says several human beings have belief in Astrology.

B - incorrect - this choice is a contradiction of the contrapositive.
   ~ postpone tasks → stars and planets are positioned to their liking (contrapositive)
   ~ postpone tasks → ~ stars and planets are positioned to their liking (contradiction of the contrapositive)

C - correct - this choice is the contrapositive of the given conditional and is therefore a valid inference.

Answers
© Gift Of Logic, Inc * Copying prohibited

9	MUST BE TRUE

The Jurassic age, when dinosaurs roamed..

The statements above, if true, will help us to reach which one of the following conclusions?
A) Dinosaurs roamed the earth 135 million years ago.
B) Dinosaurs roamed the earth since 240 million years ago.
C) Dinosaurs roamed the earth 150 million years ago.

ANSWER

Answer: C

The Jurassic period started from 238 million years ago and ended at 138 million years ago. This can be represented as 238 ↔ 138. Note that the years decrease from past to recent. 240 and 135 are before and after this period respectively.

A - incorrect -we cannot infer this because 135 million years ago is after 138 million years ago when dinosaurs became extinct on earth.

B - incorrect -since 240 million years ago is before the Jurassic period and we don't have evidence of Dinosaurs living during this period.

C - correct answer -150 million years ago falls within this range and so we can infer this.

Answers

10	MUST BE TRUE

If you purchase something with a credit card..

If the above facts are true, then which one of the following must be true?
A) If one wants to purchase a TV for 500 dollars, but has only 50 dollars in the bank, a credit card can be used.
B) If one wants to purchase a watch for 50 dollars, and has only 50 dollars in the bank, a debit card can be used.
C) If one's credit limit is 1000 dollars, he can buy a TV for 1500 dollars using a credit card.

ANSWER

Answer: B

credit card → purchase up to credit limit
debit card → must have enough money in the bank

A - incorrect - we don't know what the credit limit is and if it is more than 500 dollars – so we can't say for sure if a TV can be purchased with a credit card.

B - correct – this must be true as there is enough money in the bank to buy a watch.

C - incorrect - obviously this cannot be true because it exceeds the credit limit.

| 11 | MUST BE TRUE |

Birds can be broadly classified as either carnivorous..

The above premises support which one of the following?
A)   Herbivorous birds can eat the food that carnivorous birds eat.
B)   Carnivorous birds can eat the food that herbivorous birds eat.
C)   There is no food that both carnivorous and herbivorous birds can eat.

ANSWER

Answer: B

A simple diagram showing the classification will help in answering the question. See the diagram below.

A - incorrect - herbivorous birds can eat plants only - they cannot eat animals that the carnivorous birds eat.

B - correct - carnivorous birds can eat anything that herbivorous birds eat.

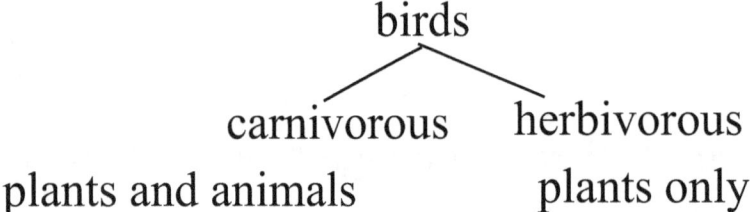

C - incorrect - plant food can be eaten by both herbivorous and carnivorous birds.

12	MUST BE TRUE

Cholesterol will increase the viscosity of blood..

Which one of the following can be properly inferred from the paragraph?
A) Dairy products and oil will impede the flow of blood.
B) Dairy products and oil will assist in the smooth flow of blood.

ANSWER

Answer: A

      if product is dairy or oil → high cholesterol in them
      cholesterol makes it difficult for blood to flow smoothly
      so, if product is dairy or oil → blood will not flow smoothly

A - correct – we can infer this from the facts presented.
B - incorrect - we cannot infer this from the facts presented.

13	MUST BE TRUE

It is unethical to seek favors of any kind from..

Which one of the following can be inferred from the above?
A) It is ethical to seek financial favors from friends who are government officials.
B) It is not ethical to seek financial favors from relatives who are not government officials.
C) It is ethical to seek favors from friends who are not government officials.

ANSWER

Answer: C

Note carefully, that the passage speaks of "favors" in general, but some of the answer choices speak of "financial favors", which is a specific kind of favor. Also, note that the answer choice talks about friends and relatives. Representing the information in a concise fashion helps to sort out the details.

Also, note how the words "ethical" and "unethical" are used. "Not ethical" is the same as unethical and "not unethical" is the same as ethical.

The information can be arranged as follows to avoid any confusion.
1) Asking favors of any kind from Government officials: not ethical
2) Asking favors from people who are not govt. officials: ethical

A - incorrect - first statement says "it is unethical to seek favors of any kind.. ". So, it is unethical to seek financial favors as well.

B - incorrect - second statement says " it is ethical to seek favors from people who are not government officials". This answer choice says that this is not ethical and so, it cannot be true.

C - correct - Applying this choice against the second statement leads us to infer that this choice must be true. It is ethical to ask for favors from friends who are not government officials.

14 MUST BE TRUE

Daniel prepared very well for the Logic exam..

The statements above, if true can lead to which one of the following conclusions?
A) Rachel is more intelligent than Daniel.
B) Daniel is more intelligent than Rachel.
C) Rachel slept more hours than Daniel the night before the exam.
D) Daniel slept more hours than Rachel the night before the exam.

ANSWER

Answer: C

Represent the information symbolically as follows:
 more sleep => higher score
 less sleep => lower score
 Rachel scored more than Daniel
 So, Rachel had more sleep than Daniel

A - incorrect - no information is available to conclude that Rachel is more intelligent than Daniel.

B - incorrect - no information is available to conclude that Daniel is more intelligent than Rachel.

C - correct - since Rachel scored more than Daniel, she would have slept more than Daniel the night before.

D - incorrect - if Daniel had slept more than Rachel, he would have scored better than Rachel.

15	MUST BE TRUE

Notwithstanding the fact that Bob is a talented sportsman..

Which one of the following can be inferred from the statement above?
A) If a sportsman is very talented he can be included in the team even if his behavior is bad.
B) Bad behavior by people other than sportsmen and sportswomen is acceptable to the younger generation.
C) Behavior that sets a bad example to youngsters will disqualify a talented sportsman from the team.

ANSWER

Answer: C

1) Bad behavior => (implies) no place in team
2) Bad behavior, particularly by sportsmen = bad example to youngsters
So, bad example to youngsters => (implies) no place in team

A - incorrect – this is exactly what the premises do not state.

B - incorrect – note the word "particularly" in the premise – this implies that bad behavior by all people (even those that are not sportsmen) do not set a good example. You could rephrase this premise as " Bad behavior by anyone, particularly by sportsmen, do not set a good example to the youngsters".

C - correct- this inference can be made from the premises.
   bad example to youngsters => (implies) no place in team (same as disqualification from the team)

16	MUST BE TRUE

In spite of repeated warnings to comply with traffic rules..

Which one of the following can be inferred from the statement above?
A) Zac has to pay an inexpensive fine.
B) Zac has to pay an expensive fine.
C) Zac does not owe any fine.

ANSWER

Answer: B

> Zac drove above speed limits.
> So he was fined.
> Penalty for speeding is expensive.
> So, Zac has to pay an expensive penalty.

A - incorrect - we cannot infer this from the premises.

B - correct – from the facts above we can infer that Zac has to pay an expensive fine.

C - incorrect - we cannot infer this from the premises.

Answers

© Gift Of Logic, Inc * Copying prohibited

17 MUST BE TRUE

Few students from Hopkins School..

Which one of the following conclusions can be made from the passage?
A) Martha and Steve are students of Preston Middle School.
B) Only two students were selected for the trip to the Space Center.
C) Steve and Anne do not go to the same school.

ANSWER

Answer: C

A - incorrect - we know that Steve is a student of Preston Middle, but we don't know to which school Martha belongs from the information available.

B - incorrect - we know from the passage, that Martha and Steve enjoyed their meeting with the astronauts and hence they were selected. We also know that a few students from each school were selected, but we don't know if only two were selected. So, this conclusion cannot be made.

C – correct - we know that Steve goes to Preston Middle School and Anne is a student of Hopkins Middle School and therefore, we can infer that they do not go to the same school.

| 1 | CANNOT BE TRUE |

The Sharks soccer team played five games..

If the above facts are true, which one of the following cannot be true?
A) The Sharks team won five games.
B) The Sharks team lost three games and the Panthers team lost two games.
C) The Sharks team won two games and lost two games against the Panthers.

ANSWER

Answer: C

A - incorrect - this can be true if the Sharks won all the five games. We cannot conclude that this choice cannot be true.

B - incorrect - this can be true – since five games were played and all of them is either a win or a lose, there should be a total of five wins and five losses, for both teams combined. This choice indicates that a total of five games were lost, which is a possibility.

C - correct - this cannot be true - if Sharks won two games, then they should have lost three games and not two as this choice indicates.

2   CANNOT BE TRUE

Students in a school must choose courses..

If the above conditions are true, then which one of the following cannot be true regarding Jennifer, a student of the school?
A) She takes Algebra, Trigonometry, and Calculus.
B) She takes Trigonometry and Calculus, but not Algebra.
C) She takes Algebra and Calculus, but not Trigonometry.

ANSWER

Answer: C

Note the conditions imposed for taking courses. If you take Algebra, you must take Trigonometry. If you take Trigonometry, you must take Calculus.

       Algebra → Trigonometry (A → T)
       Trigonometry → Calculus (T → C)
       Therefore, Algebra → Calculus (A → C)

A - incorrect – this must be true - if Jennifer takes Algebra, she must take Trigonometry, which will force her to take Calculus.

B - incorrect – this can be true - if she takes Trigonometry, she must take Calculus which she does – but, she does not take Algebra which is ok because the rules don't say that if she takes Trigonometry, she should take Algebra. The converse of Algebra → Trigonometry is not true.

C - correct – this cannot be true – she takes Algebra, but not Trigonometry which is a violation of the Algebra → Trigonometry condition. If she takes Algebra, she must take Trigonometry.

3 — CANNOT BE TRUE

Calcium is very important for bones..

If the above information is true, which one of the following must be false?
A) If a person does not have sufficient calcium in his diet, he will have bone mass below the normal level.
B) If a person does not have below normal level of bone mass, then he does not have sufficient calcium in his diet.
C) If a person has a normal bone mass level, he will have sufficient calcium in his diet.

ANSWER

Answer: B
condition: ~sufficient calcium → bone mass below normal level
contrapositive: ~bone mass below normal → sufficient calcium

A - incorrect – this is just a restatement of the premise. This statement is true.

B - correct – this cannot be true. This choice violates the contrapositive condition, as can be seen by comparing them below.

   choice: ~bone mass below normal → ~sufficient calcium
   contrapositive: ~bone mass below normal → sufficient calcium

C - incorrect – this is the contrapositive statement and is true.
       contrapositive: ~ bone mass below normal → sufficient calcium

Note that the person has a normal bone mass level. So, the person satisfies the antecedent, namely ~ bone mass below normal. So, we can infer that he will have sufficient calcium.

4   CANNOT BE TRUE

A car dealer sold 100 cars during the first month..

If the above facts are true, which of the following cannot be true?
A) The dealer sold more cars in March 2006 than he did in March 2007.
B) Excluding the first month, more cars were sold in 2007 than in 2006.
C) The dealer sold the same number of cars during both years.

ANSWER

Answer: B

The information can be transcribed as follows.
    2006 - (100)Jan+ (900)other =1000
    2007 - (900)Jan+ (100)other =1000

A - incorrect – sufficient information is not available for the month of March of 2006 and 2007 to decide if this statement is true or false.

B - correct – this cannot be true. In 2006, 900 cars were sold in all months except the first month, whereas only 100 cars were sold in 2007 for the same period.

C - incorrect – this answer choice is true.

5 CANNOT BE TRUE

P, Q and R are names of three groups of people..
If the above facts are true, which of the following cannot be true?
A) No one belongs to P, Q and R.
B) No one belongs to both P and R.

ANSWER

Answer: B

A - incorrect - must be true - See the Venn diagram. The area covered by the three groups is shaded. This means that no one can belong to these three groups.

B - correct - cannot be true - if this were to be true, this area would be blacked out after drawing the premises.

No one belongs to both P and Q
No one belongs to both Q and R

No one belongs to both P and R - cannot be true 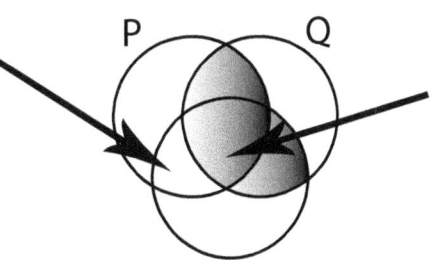 No one belongs to P, Q and R - true

No one belongs to both P and Q
No one belongs to both Q and R

Answers
© Gift Of Logic, Inc * Copying prohibited

6 CANNOT BE TRUE

To create a vacuum in a given area, air inside the area..

If the above statements are true, which one of the following must be false?
A) In a thermos flask with an unbroken vacuum, hot water will remain hot and cold water will remain cold.
B) A thermos flask whose vacuum is broken will retain energy as efficiently as a thermos flask with an unbroken vacuum.

ANSWER

Answer: B

    vacuum does not have air in it
     energy cannot transfer through vacuum
      if vacuum is broken, air will rush to fill the area

A - incorrect - this statement is true. Since the vacuum is unbroken, energy will not flow in or out of the thermos flask.

B - correct – this cannot be true – since the vacuum is broken, per the given facts, air will rush to fill the vacuum. This will allow the energy inside the flask to travel out of the flask. So, a thermos flask whose vacuum is broken will not retain energy as efficiently as one with an unbroken vacuum.

Answers

7   CANNOT BE TRUE

There are five buildings of type A and ten buildings..

If the above statements are true, which one of the following must be false?
A) After applying the blue paint on half the buildings of type B, there are the same number of buildings of each type.
B) There is at least one building of type C after the green paint is applied to half the buildings of type B.

ANSWER

Answer: B

    5A, 10 B
B → blue paint → C
B → green paint → A

A – incorrect – this is true – after applying the blue paint on five blue buildings, we have 5A, 5B and 5C buildings left.

B – correct – this cannot be true – after green paint is applied to half of the buildings of type B, we have 10 A(5A+5A), 5 B and 0 C left. There is no building of type C available.

Answers
© Gift Of Logic, Inc * Copying prohibited

| 8 | CANNOT BE TRUE |

To teach Physics at any University..

Which one of the following is an incorrect conclusion derived from the above facts?
A) Mr. Rogers has a Ph.D in Physics.
B) If you do not have a Ph.D in Physics, you cannot teach Physics at any University.

ANSWER

Answer: A

> teach Physics at any Univ → must have Ph.D in Physics
> Rogers teaches Physics at Everest middle school.

A - correct – Mr. Rogers teaches Physics at a middle school, not at an University. So, we cannot conclude that he has a Ph.D in Physics with this information.

B - incorrect - this is the contrapositive of the given condition, and so this must be true.

> teach Physics at any Univ → must have Ph.D in Physics
> ~Ph.D in Physics → ~teach Physics at any University

answers
© Gift Of Logic, Inc * Copying prohibited

| 9 | CANNOT BE TRUE |

Mold, a type of fungus, is a black or green hairy mass..

If the information in the above paragraph is true, which one of the following cannot be true?
A) Fungus can be found inside the human body.
B) Since fungus is destructive in nature, it is of no use to mankind.
C) When fungus settles on organic matter, the matter disintegrates into its components.

ANSWER

Answer: B

Note the use of the counter premise "However".

A - incorrect – this is a true premise.

B - correct – this cannot be true. The premise only says that the substances that it infects are of no use to mankind, but not the fungus itself. In fact, the last sentence points out that the fungus performs an essential function, thereby meaning that it is indeed useful to mankind.

C - incorrect – this is true – as stated in the last premise, it breaks down organic matter.

| 10 | CANNOT BE TRUE |

Gastric juice is a fluid created in the stomach..

If the information presented above is true, which one of the following must be false?
A) Smoking is the only cause of ulcer in stomach.
B) Hydrochloric acid is required to break down the food in the stomach.

ANSWER

Answer: B
Carefully note the cause and effect described in the passage.
c→ is the symbol for cause and effect. Refer to the Primer for discussion on causal relationships.

HCA = hydrochloric acid
Food is broken down by Gastric Juice.
Gastric Juice is made up of HCA, Pepsin, Mucin
Mucin protects wall of stomach
No (mucin) protection c→ ulcer
High level of HCA, smoking, stress c→ erosion of stomach walls (no protection)
So, high level of HCA, smoking, stress c→ ulcer

A – correct – cannot be true- high level of HCA, smoking and stress all cause an erosion of stomach wall leading to ulcer. So, smoking is not the only cause of stomach ulcer.

B – incorrect – this is true – since food is broken down by gastric juice and since gastric juice contains HCA, it must be true that HCA is required to break the food down in the stomach.

Answers

| 11 | CANNOT BE TRUE |

Many women in China like to assume names that signify..

If the above statements are true, which one of the following must be false?
A) "Lanfen", which in Chinese means "orchid fragrance" is a name that some Chinese women would not like to be called with.
B) Women in China do not like to have their name associated with anything fragrant.

ANSWER

Answer: B

A - incorrect – can be true – it is possible that some Chinese women would like to be called as Lanfen, but information is not sufficient to infer that this is true or false.

B - correct – this cannot be true –as survey results have shown, at least twenty percent of women in each city have a name that implies fragrance. This choice says, women, meaning all women in China do not like to have a name associated with fragrance, which clearly cannot be true given the facts.

12	CANNOT BE TRUE

The number "3" is a lucky number..

If the statement above is true, which one of the following cannot be true?
A) A player with jersey# 3 did not play and yet Sharks won a game.
B) Sharks did not win the game even though Jersey # 3 played.
C) The Panthers had a player with jersey# 3 and they won the game.

ANSWER

Answer: B

The condition is:
    person wearing jersey#3 plays → win the game.

A - incorrect – this statement can be true – the passage says that if jersey #3 plays, Sharks will win – this means that Sharks can win even if jersey#3 does not play. Careful understanding of the condition is required. The condition "If P then Q" allows Q to happen without P. Refer to the Primer for detailed discussion on conditional reasoning.

B - correct – this statement cannot be true – if jersey#3 plays for Sharks, they will surely win according to the condition. To say that they did not win is a violation of the condition and cannot be a true statement.

C - incorrect – may be true - it is possible that Panthers had a player with jersey#3 and that the Panthers won. This fact cannot be inferred from the given facts.

13 CANNOT BE TRUE

Dr. Henry liked solitude.

If the above information is true, then which one of the following cannot be true?

A) Dr. Henry's discoveries in physics and chemistry were performed when no one except himself was present.
B) Dr. Henry's team made several scientific discoveries in physics and chemistry.
C) Dr. Henry made outstanding discoveries in physics and chemistry during his time.

ANSWER

Answer: B

He performed "all" his experiments without anyone's help.
He made several discoveries in Physics and Chemistry.
So, he made several discoveries in Physics and Chemistry without anyone's help.

A - incorrect - this is a true statement. He performed his experiments in solitude.

B - correct - this cannot be true. Dr. Henry did all his experiments without anyone's help. So, he did not have a team.

C - incorrect - this is true - as stated in the premise. Stunning discovery is the same as outstanding discovery.

14	CANNOT BE TRUE

The Stanley Middle School starts sharp at 8 AM.

If the above statements are true, then which one of the following cannot be true?

A) More students were tardy to school during the previous year when compared to last year.
B) A higher percentage of students were tardy during the previous year when compared to last year.
C) There were more punctual students during the previous year when compared to last year.

ANSWER

Answer: C

last year - 50 tardy, total 500 students
previous year - 100 tardy, 200 students

A - incorrect - this is true - more number of students (100) were tardy during the previous year when compared to last year (50).

B - incorrect - this is true - 50 percent of the students were tardy during the previous year, whereas only 10 percent was tardy last year.

C - correct - this cannot be true - there were more tardy students during the previous year compared to last year. This means that there were less number of punctual students during the previous year compared to last year.

| 15 | CANNOT BE TRUE |

Everyone knows how to handle..

Which one of the following does not follow from the facts presented above?
A) Off-the-shelf analgesics can get rid of all types of headaches.
B) Headaches caused by migraine cannot be relieved by off-the-shelf analgesics.

ANSWER

Answer: A

Note that most headaches can be relieved by off-the-shelf analgesics, but headaches caused by migraine need prescription medicine, thereby implying that off-the-shelf analgesics are not effective in treating headaches caused by migraine.

A - correct - cannot be true - off-the-shelf analgesics are not effective in treating headaches caused by migraine.

B - incorrect - this is true - that is why prescription medicine is required to handle headaches caused by migraine.

16 CANNOT BE TRUE

Employees of Alpha Construction Company..

Which one of the following does not follow from the facts presented above?

A) When a disaster strikes the city where the company is located, the employees can claim benefits from the disaster relief fund.
B) When a disaster strikes the city where the company is located, anyone can claim benefits from the fund.
C) If you are not a beneficiary, then you did not contribute to the fund.

ANSWER

Answer: B

A - incorrect - this is true. This is what the fund is set up for.

B - correct - this cannot be true. The rule states that only those who contributed to the fund can be beneficiaries. So, anyone cannot claim benefits from the fund.

C - incorrect - this is true. The given fact is that only those employees who contributed to the fund can claim benefits from the fund.

    contribution → beneficiary

The contrapositive of this statement is:
    ~beneficiary → ~contribution

Choice C describes the contrapositive of the condition and so, it is true.

Answers
© Gift Of Logic, Inc * Copying prohibited

| 17 | CANNOT BE TRUE |

Calls to the Customer Service department..

If the information in the above passage is true, then which one of the following does not follow?
A) The delays have worsened since the complaints were made.
B) The delays have improved since the complaints were made.

ANSWER

Answer: A

A - correct - this cannot be true - after the complaints were made, the management rectified the situation immediately. So, the delays could not have worsened after the complaints.

B - incorrect - this is true - after the complaints were made, the management rectified the situation immediately. So, the delays would have improved.

SUDOKU

Solve the following Sudoku. A correctly solved Sudoku has numbers 1-9 appearing only once in each row, each column and each 3x3 grid. Solving Sudokus will help you to gain valuable analytic skills.

6	1	9	8	3	7	4	2	5
3	4	5	2	1	6	8	7	9
7	2	8	5	4	9	1	3	6
4	3	7	1	9	8	5	6	2
1	8	6	7	5	2	9	4	3
5	9	2	3	6	4	7	8	1
2	7	1	6	8	5	3	9	4
9	6	3	4	7	1	2	5	8
8	4	5	9	2	3	6	1	7

Answers
© Gift Of Logic, Inc * Copying prohibited

SUDOKU

Solve the following Sudoku. A correctly solved Sudoku has numbers 1-9 appearing only once in each row, each column and each 3x3 grid. Solving Sudokus will help you to gain valuable analytic skills.

3	7	2	8	1	6	5	9	4
6	4	9	3	7	5	8	1	2
1	8	5	4	9	2	3	6	7
5	2	6	7	8	3	9	4	1
4	9	8	5	2	1	6	7	3
7	3	1	6	4	9	2	8	5
2	6	4	9	5	7	1	3	8
9	1	7	2	3	8	4	5	6
8	5	3	1	6	4	7	2	9

Answers

3 SUDOKU

Solve the following Sudoku. A correctly solved Sudoku has numbers 1-9 appearing only once in each row, each column and each 3x3 grid. Solving Sudokus will help you to gain valuable analytic skills.

1	8	5	7	9	4	3	6	2
6	2	3	8	5	1	9	7	4
9	7	4	2	3	6	5	8	1
2	4	8	3	1	7	6	9	5
3	5	9	6	2	8	1	4	7
7	6	1	9	4	5	2	3	8
8	1	2	4	6	3	7	5	9
5	3	7	1	8	9	4	2	6
4	9	6	5	7	2	8	1	3

Answers

© Gift Of Logic, Inc * Copying prohibited

1	POSITIONING

In the empty cells, write the letter Y so that a X is not below a Y.

	X	
X	Y	
Y	Y	X

1) How many Y can you place in the empty cells?
Answer: 3, as shown above.

2	POSITIONING

In the empty cells, draw a circle so that there is no square to its left.

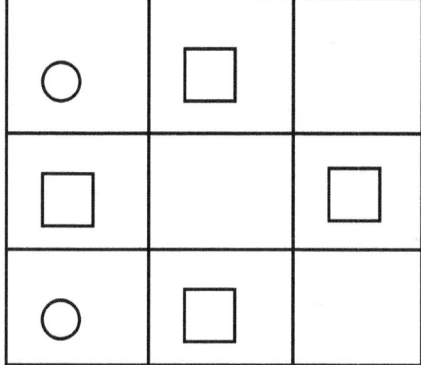

1) How many circles can you place in the empty cells?
Answer: 2, as shown above.

Answers
© Gift Of Logic, Inc * Copying prohibited

3	POSITIONING

A string is tied..

1) Can the bird sit on the pole 1? Yes

2) In how many positions can the bird sit? 3 (1, 3 and 5, the odd numbered positions)

4	POSITIONING

There are several slots..

1) A 40 Watt slot can hold a 60 Watt bulb.
Answer: B) False. A 40 Watt slot can hold only a 40 Watt bulb.

2) The number of slots where you can plug a 60 Watt bulb is
Answer: A) 3, in the three 60 Watt slots.

3) The number of slots where you can fit a 40 Watt bulb is
Answer: B) 5 A 40 Watt bulb can be plugged into a 40 Watt slot or a 60 Watt slot. There are two 40 Watt slots and three 60 Watt slots.

5	POSITIONING

The grid above shows..

The sequence of moves that the Horse can make to reach the Elephant is
  Answer: B) (1,2,5,8) or (1,4,7,8)

Answers
© Gift Of Logic, Inc * Copying prohibited

6	POSITIONING

The grid above shows..

The sequence of moves that the Horse can make to reach the monkey is
Answer: C) (2,1,4,7) and (7,8,9,6)

7	POSITIONING

The grid above shows..
Is it possible to fill the blank cells with numbers 1 and 2 so that each row and each column has two different numbers?

Answer: B) No. If each row has different numbers then each column will end up having the same numbers.

1	2
1	2

8	POSITIONING

3	1	2
1	2	3
2	3	1

Answers
© Gift Of Logic, Inc * Copying prohibited

9	POSITIONING

Fill the blank rows with symbols △, ○, and □ so that each row and each column has one of each shape.

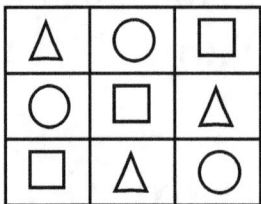

10	POSITIONING

There are nine train stations..

At which station should they get down so that they can meet each other?
Answer: B) station# 7

Train A will stop at stations 1, 3, 5, 7 and 9. Train B will stop at 1, 4, and 7. So, if both Hansel and Gretel get down at station# 7, they can meet.

11	POSITIONING

RB	LRB	LB
RAB	LRAB	LAB
RA	LRA	LA

12 POSITIONING

Place the letters A, B, C, D, E ,F, G and H..

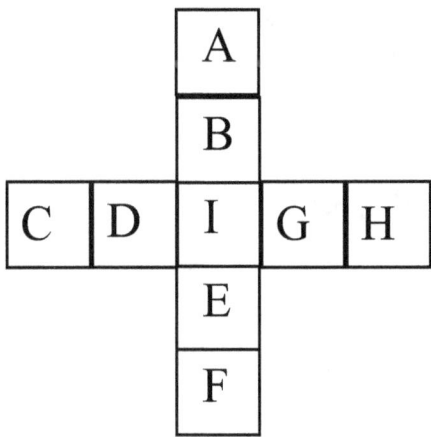

13 POSITIONING

A	H	G
B	C	F
I	D	E

Place the letters A, B, C, D, E, F, G, H, and I in the grid above. The letter A must be in the first cell (top-left) and must be immediately above B. C must be immediately to the right of B. F must be to the right of C, below G , and immediately above E. I must be below B and D must be immediately after I.

Answers

14	POSITIONING

I	H	G
F	E	D
C	B	A

15	POSITIONING

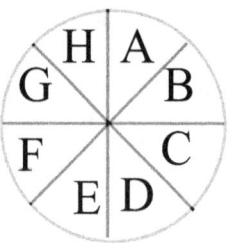

16	POSITIONING

A		
2		C
	B	

The empty cells in the grid shown above..
1) What is the maximum number of letters that the grid can hold?
    Answer: 8

We need at most three numbers, but the grid already has one number (2). So, we do not need any more numbers since we need to maximize the number of letters in the grid. At most three numbers does not mean that we need three numbers. We can stop with the one number that is already there (2). So, the grid can hold 9-1=8 letters, including the three letters that are already there.

Answers

17 POSITIONING

Red, Green, Blue, Orange balls are .. R,G,B,O,Y

Red ball must be the first ball. >> R1
Green and Blue balls must be together. >> GB ∦ BG (∦ is the symbol for OR)
Orange ball must be immediately before the green ball. >> OG (so, BG is eliminated as a possibility). Also, OG and GB can be chained as OGB.
Yellow ball can be placed before or after the orange ball. >> O-Y ∦ Y-O

1) Green ball can be the third ball. Answer: A) True

R,G,B,O,Y
R1
OGB
O-Y ∦ Y-O

1	2	3	4	5
R	O	G	B	Y

If Green is the third ball, then rule OGB means that O must be the second ball and B must be the fourth ball. Rule O-Y means that Y comes after O in the last position. So, all the balls can be placed as shown above without violating any rules.

2) Yellow ball can be the third ball. Answer: B) False

R,G,B,O,Y
R1
OGB
O-Y ∦ Y-O

1	2	3	4	5
R		Y		

If Yellow is the third ball, then OGB cannot be placed together.

1	GROUPING

A tray contains alphabets..What are the possible picks?   Write them below.
If A is picked, D cannot be picked. So A and D cannot be in the selection. But,
D can be in the selection with B or C.
    AB, AC, BC, BD, CD

2	GROUPING

Pick two fruits from the tray above. The condition is that if apple is picked, then grape must also be picked. This can be represented in symbolic as A → G. Note that the converse G → A is not true. Refer to Primer for a discussion on conditional reasoning.

What are the possible picks?   Write them below.

Regardless of the rules, any two fruits can be selected as follows:
    apple,orange   orange,grape   grape, apple

Because of the rule A →G, the first choice (apple, orange) cannot be selected since grape is missing. So, the valid selections are (orange,grape) and (grape, apple). Note that grape can be picked without picking apple.

3	GROUPING

Select one alphabet..

What are the possible picks? Write them below.
    A,1   A,2   B,1   B,2

Answers

144

© Gift Of Logic, Inc * Copying prohibited

4	GROUPING

There are five Science books.. You must select at least 2 Science books, at most 2 Math books and at least 3 history books.

1) What is the maximum number of books that can be selected?
   Answer: 5 Science + 2 Math + 5 History = 12 books.

2) What is the minimum number of books that can be selected?
   Answer: 2 Science + 0 Math + 3 History = 5 books.

5	GROUPING

Team P has four members..
A cannot be selected with E or F or both. >> ~AE, ~AF, ~AEF

1) Which of the following selections are valid?
   A)   A   B   G   F >> invalid, if A is selected, F must not be selected
   B)   E   H   A   D >> invalid, if A is selected, E must not be selected
   C)   C   B   E   F >> valid

6	GROUPING

Team A (A,B,C,D) and team B (E,F,G,H).
If A is selected, E must be selected. A → E
If C is selected, G must be selected. C → G

1) Which of the following selections are valid?
A)   A   B   F   G >> violates A → E rule
B)   C   A   G   E >> valid
C)   E   G   B   D >> valid

Answers

7	GROUPING

Class A has 7 boys..

To make the classes to be of same size, what must be done?

Answer: A) Move 1 boy from class A to class B. This will make the class size be 10 in both classes.

8	GROUPING

Box-1 contains..

How many squares and rectangles are there in the regrouped boxes?

    Box-1 5 Squares;4 triangles.   Box-2: 5 Squares;4 triangles

9	GROUPING

From a group of 5 cars..

How many groups can be selected?   2 groups as shown below.

5 vehicles need to be selected, there should be less number of cars than trucks, and there should be at least one vehicle of each type in the selection.

    1 car, 4 trucks

    2 cars, 3 trucks

10	GROUPING

From a basket of 5 apples and 5 oranges..

5 fruits must be selected, there should be more apples than oranges, and there should be at least one fruit of each type.

How many groups can be selected?

    4 apples, 1 orange

    3 apples, 2 oranges

Answers

PATTERN PERCEPTION

Question#	Answer
1	A
2	B
3	A
4	A
5	B
6	A
7	B
8	A

FIGURE FORMATION

Question#	Answer
1	B
2	A
3	B
4	A
5	A
6	B
7	B

PAPER FOLDING AND CUTTING

Question#	Answer
1	A
2	C
3	B

FIGURE MATRIX

Question#	Answer	Reasoning
1	B	sports related materials
2	A	reading materials
3	B	domestic pets
4	A	long distance transportation
5	B	policeman uses handcuffs; doctor uses stethoscope
6	B	cat eats mouse; bird eats earthworm
7	A	candle is used when bulb does not work; hand fan is used when electric fan does not work
8	B	vase holds flowers; cabinet holds books

RULE DETECTION

Question#	Answer
1	A
2	A
3	B
4	B
5	A
6	B

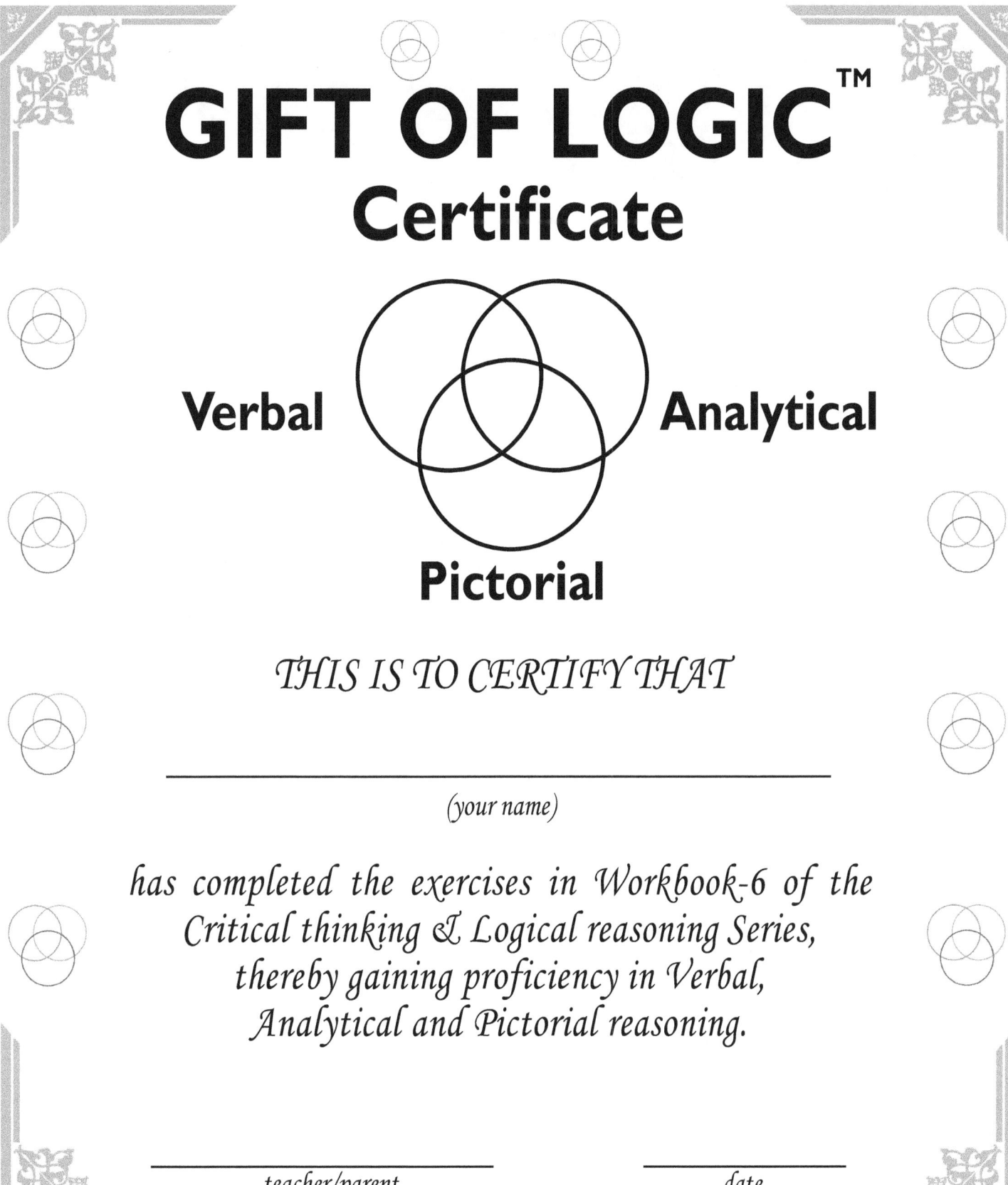

NOTES

www.ingramcontent.com/pod-product-compliance
Lightning Source LLC
Chambersburg PA
CBHW080253180526
45167CB00006B/2510